Modern Methods of Plant Analysis

New Series Volume 2

Editors
H.F. Linskens, Nijmegen
J.F. Jackson, Adelaide

Volumes Already Published in this Series:

Volume 1: Cell Components
1985, ISBN 3-540-15822-7

Volume 2: Nuclear Magnetic Resonance
1986, ISBN 3-540-15910-X

Nuclear Magnetic Resonance

Edited by
H. F. Linskens and J. F. Jackson

Contributors
C. Abell P. Bendel C. Dumas R. C. Hirst C. Kerhoas
D. C. McCain J. K. M. Roberts T. J. Simpson J. Visintainer

With 69 Figures

Springer-Verlag
Berlin Heidelberg New York Tokyo

Professor Dr. HANS-FERDINAND LINSKENS
Botanisch Laboratorium
Faculteit der Wiskunde en Natuurwetenschappen
Katholieke Universiteit
Toernooiveld
NL-6525 ED Nijmegen
The Netherlands

Professor Dr. JOHN F. JACKSON
Department of Biochemistry
Waite Agricultural Research Institute
University of Adelaide
Glen Osmond, S.A. 5064
Australia

ISBN-13:978-3-642-82611-5 e-ISBN-13:978-3-642-82609-2
DOI: 10.1007/978-3-642-82609-2

Library of Congress Cataloging-in-Publication Data. Main entry under title: Nuclear magnetic resonance. (Modern methods of plant analysis; new ser., v. 2) Includes bibliographies. 1. Plants – Analysis. 2. Nuclear magnetic resonance. I. Linskens, H.F. (Hans F.), 1921–. II. Jackson, J.F. (John F.), 1935–. III. Abell, C. IV. Series. QK865.N8 1986 581'.028 85-31734
ISBN-13:978-3-642-82611-5 (U.S.)

2131/3130-543210

Introduction

Modern Methods of Plant Analysis

When the handbook *Modern Methods of Plant Analysis* was first introduced in 1954 the considerations were:

1. the dependence of scientific progress in biology on the improvement of existing and the introduction of new methods;
2. the difficulty in finding many new analytical methods in specialized journals which are normally not accessible to experimental plant biologists;
3. the fact that in the methods sections of papers the description of methods is frequently so compact, or even sometimes so incomplete that it is difficult to reproduce experiments.

These considerations still stand today.

The series was highly successful, seven volumes appearing between 1956 and 1964. Since there is still today a demand for the old series, the publisher has decided to resume publication of *Modern Methods of Plant Analysis*. It is hoped that the New Series will be just as acceptable to those working in plant sciences and related fields as the early volumes undoubtedly were. It is difficult to single out the major reasons for success of any publication, but we believe that the methods published in the first series were up-to-date at the time and presented in a way that made description, as applied to plant material, complete in itself with little need to consult other publications.

Contributing authors have attempted to follow these guidelines in this New Series of volumes.

Editorial

The earlier series *Modern Methods of Plant Analysis* was initiated by Michel V. Tracey, at that time in Rothamsted, later in Sydney, and by the late Karl Paech (1910–1955), at that time at Tübingen. The New Series will be edited by Paech's successor H.F. Linskens (Nijmegen, The Netherlands) and John F. Jackson (Adelaide, South Australia). As were the earlier editors, we are convinced "that there is a real need for a collection of reliable up-to-date methods for plant analysis in large areas of applied biology ranging from agriculture and horticultural experiment stations to pharmaceutical and technical institutes concerned with raw material of plant origin". The recent developments in the fields of plant biotechnology and genetic engineering make it even more important for workers in the plant sciences to become acquainted with the more sophisticated methods, which sometimes come from biochemistry and biophysics, but which also have been developed in commercial firms, space science laboratories, non-university research institutes, and medical establishments.

Concept of the New Series

Many methods described in the biochemical, biophysical, and medical literature cannot be applied directly to plant material because of the special cell structure, surrounded by a tough cell wall, and the general lack of knowledge of the specific behavior of plant raw material during extraction procedures. Therefore all authors of this New Series have been chosen because of their special experience with handling plant material, resulting in the adaptation of methods to problems of plant metabolism. Nevertheless, each particular material from a plant species may require some modification of described methods and usual techniques. The methods are described critically, with hints as to their limitations. In general it will be possible to adapt the methods described to the specific needs of the users of this series, but nevertheless references have been made to the original papers and authors. While the editors have worked to plan in this New Series and made efforts to ensure that the aims and general layout of the contributions are within the general guidelines indicated above, we have tried not to interfere too much with the personal style of each author.

Volume Two – NMR Spectroscopy

The first volume in the New Series dealt with Cell Components and Organelles, and was planned to be useful as groundwork for later volumes. This second volume concentrates on a powerful technique, Nuclear Magnetic Resonance (NMR) spectroscopy, which is now beginning to be applied to plant studies in a major way. We believe that this is the first time that a series of chapters dealing with the application of NMR to plant cells has been gathered together in one volume. The authors chosen are some of the leading exponents of this technique in their respective fields, and they cover the use of ^{13}C, proton, deuterium, and ^{31}P in NMR investigations. As far as plant studies are concerned, the volume covers such diverse topics as general metabolism in higher plants and fungi, intercellular pH, energy status, DNA structure analysis, multiple quantum two-dimensional NMR spectroscopy, chloroplast orientation, rubber analysis, and the use of NMR for pollen viability. We feel that the authors have all together given an excellent guide to the possibilities of NMR applications to the plant sciences, and we hope that this volume gives more impetus to its use in the future.

Acknowledgments. The editors express their gratitude for the excellent cooperation with the publisher, with Dr. Dieter Czeschlik and especially with Mrs. Linda Teppert. We are grateful also to the invaluable assistance of Mrs. Jose Smits-Broekmans.

Nijmegen and Adelaide, February 1986

H. F. LINSKENS
J. F. JACKSON

Contents

^{13}C-NMR in Metabolic Studies
T. J. SIMPSON (With 15 Figures)

Determination of the Energy Status of Plant Cells by ^{31}P-Nuclear Magnetic Resonance Spectroscopy
J. K. M. ROBERTS (With 8 Figures)

The Use of NMR Spectroscopy to Follow Deuterium in Studies of Fungal Metabolism
C. ABELL (With 2 Figures)

List of Contributors

ABELL, CHRISTOPHER, University Chemical Laboratory, Lensfield Road, Cambridge CB2 1EW/United Kingdom

BENDEL, PETER, School of Chemistry, Tel-Aviv University, Ramat-Aviv 69978, Israel

DUMAS, CHRISTIAN, Reconnaissance Cellulaire et Amélioration des Plantes, Université Claude Bernard, Lyon 1, CNRS UM 380024, F-69622 Villeurbanne Cedex/France

HIRST, R. C., The Goodyear Tire & Rubber Company, Research Division, Akron, OH 44316/USA

KERHOAS, CATHERINE, Reconnaissance Cellulaire et Amélioration des Plantes, Université Claude Bernard, Lyon 1, CNRS UM 380024, F-69622 Villeurbanne Cedex/France

MCCAIN, DOUGLAS C., Department of Chemistry, Southern Station Box 5043, University of Southern Mississippi, Hattiesburg, MS 39406/USA

ROBERTS, JUSTIN K. M., Department of Biochemistry, University of California, Riverside, CA 92521/USA

SIMPSON, THOMAS, J. Department of Chemistry, University of Edinburgh, West Mains Road, Edinburgh EH9 3JJ/United Kingdom

VISINTAINER, J., The Goodyear Tire & Rubber Company, Research Division, Akron, OH 44316/USA

^{13}C-NMR in Metabolic Studies

T. J. Simpson

1 Introduction

The study of natural products and the development of organic chemistry have always been closely related, and the rapid growth of organic chemistry in the 19th century was due largely to the efforts made in isolating and characterising some of the more biologically active plant metabolites. A lasting result of this is the predominance of the methodology of organic chemistry in the study of the biosynthesis and metabolism of secondary plant products, in contrast to the study of primary metabolism in which biochemical methods have held sway.

In this chapter the general principles involved in applying stable isotope labelling methods in conjunction with ^{13}C NMR spectroscopy to the study of secondary metabolism will be described. For reasons which will become obvious below, these techniques will be illustrated in many cases by reference to studies in microorganisms. Whether micro-organisms, in particular the lower fungi, can be regarded as plants is debatable, but as many of the techniques have been developed and applied in studies of fungal metabolism, their use for illustrative purposes is appropriate. The latter parts of the sections will survey the metabolic pathways in higher plants which have been studied using ^{13}C NMR. It is hoped to convey the basic requirements for carrying out ^{13}C-labelling studies to both potential users and those interested in the results, not only to give some idea of the strengths and advantages of these techniques but also their limitations and the difficulties and disadvantages associated with their use, and general considerations in interpreting the results.

As a necessary preamble, a brief discussion follows of some basic principles and considerations in biosynthetic studies and of radioisotope tracer methods.

2 The Design and Interpretation of Labelling Experiments

Some of the principles involved in carrying out biosynthetic experiments have been usefully discussed in a number of texts (Bu'lock 1965; Manitto 1981; Herbert 1981; Brown 1972). The most fundamental method for establishing the intermediates involved in a biosynthetic pathway is to label likely compounds with an isotope and then use this isotopic label to assess the extent to which the compounds are incorporated into the products of the pathway. A number of problems are associated with this.

Low or negligible incorporation, even of true precursors, is often observed. Compounds of general metabolic importance may show low incorporation into the metabolites of interest due to the diversity of other routes open to them. In studies using whole organisms, particularly in the higher plants, structurally complex or highly polar intermediates may not be able to cross cell membranes, or may not be transported to the site of biosynthesis in highly differentiated organisms, or they may be degraded or otherwise metabolised before they can be utilised. Therefore a negative result is always inconclusive and must be regarded with caution.

Careful interpretation is also required in the case of apparently positive results, particularly when using the isolation and incorporation of other metabolites to obtain information on the later stages of a pathway. Scheme 1 illustrates some

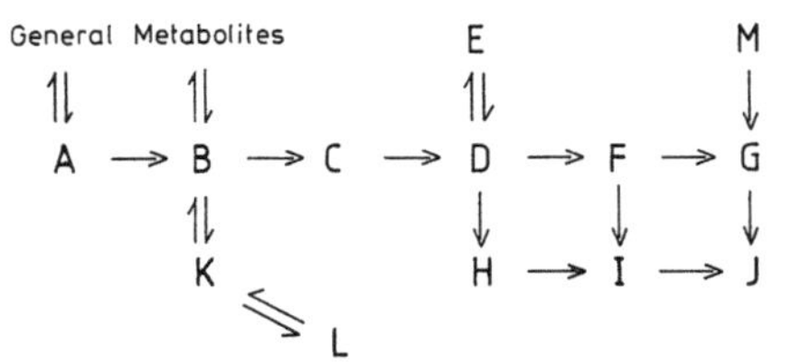

Scheme 1

J is the metabolite of interest.
A and B are precursors. (The term "prime-precursor" has been coined for the species furthest into the secondary route which is also involved in general metabolism – in this case B.)
C and D are obligatory intermediates.
F, G, H, and I are non-obligatory intermediates.
K and L are not intermediates but may label J non-specifically.
E is a "shunt metabolite".
M is not a natural intermediate but is utilisable by the organism.

of the problems. Thus, a compound (L) may be isolated, which when labelled appears to be incorporated into the final metabolite. However, unless it can be shown that the label is incorporated *specifically* (i.e. from a known position in the precursor into a corresponding position in the metabolite of interest), the possibility remains that degradation to a common early precursor followed by incorporation of this degradation product has taken place. A striking example of this was provided by the reported intact incorporation of [14-^{14}C] palmitic acid **(1)** into brefeldin A **(2)** by cultures of *Penicillium brefeldianum*. However, subsequent work (Cross and Hendley 1975) showed that the observed incorporation was due to degradation of palmitic acid to produce [2-^{14}C]-acetyl CoA **(3)** which was then incorporated to produce brefeldin A labelled as shown in Scheme 2.

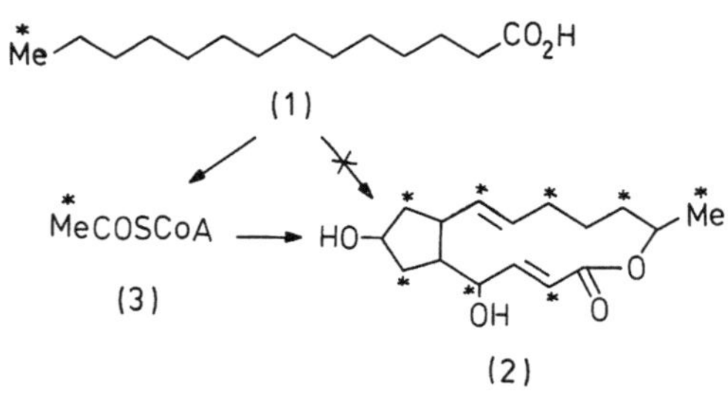

Scheme 2

Even if specific incorporation can be demonstrated, the supposed intermediate may not lie on the main pathway, but may be formed as shown in scheme 1 reversibly as a "shunt" metabolite **(E)** from a compound **(D)** which is on the pathway. A further complicating factor is the often observed lack of specificity of the enzymes of secondary metabolism giving rise to a metabolic grid, as illustrated by compounds **(D)**, **(F)**, **(G)**, **(H)**, **(I)**, and **(J)** in Scheme 1. As indicated there are a priori three possible routes from **(D)** to **(J)**, so that **(F)**, **(G)**, **(H)**, and **(I)** are all possible but not *obligatory* intermediates. An interesting example of this and of its solution has been described by Campbell in a study of mycophenolic acid **(4)** biosynthesis (Doerfler et al. 1980).

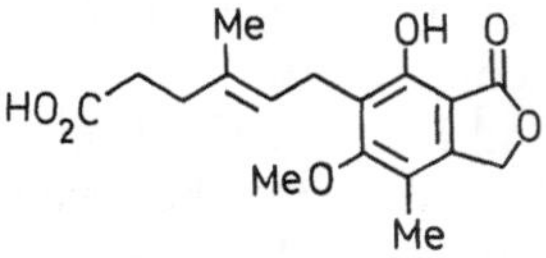

(4)

3 Radioisotope Labelling

Until the advent of ^{13}C NMR techniques, labelling with the radioactive isotopes ^{14}C and ^{3}H was the most powerful method of biosynthetic investigation. The major advantage of these radionuclides is their negligible natural abundance, which, when coupled with the low limits of detection of modern liquid scintillation counters (Faires and Boswell 1981) allows very low levels of precursor incorporation to be accurately measured. Only very small quantities of labelled precursors are required, providing that the specific activity is sufficiently high, thus the usual assumption that the addition of tracer does not itself affect the metabolic pathway under study is likely to be valid.

The main disadvantages are the need to obtain radiochemically pure compounds and the problem of establishing specificity of incorporation of label. This usually requires experimentally demanding and often impractical degradative chemistry, especially where the metabolite has been labelled from a simple precursor such as acetate or mevalonate and is labelled in many sites. More complex intermediates may be labelled in two sites with both ^{3}H and ^{14}C in a known ratio and administered to an organism. The isolation of a product with the same ^{3}H : ^{14}C ratio as the precursor provides good evidence for intact incorporation of the precursor. This remains a good method for obtaining information on the stereochemical fate of particular hydrogen atoms in a molecule, although ^{2}H, ^{13}C double labelling (see Sects. 8.1 and 8.2) and ^{2}H labelling in conjunction with ^{2}H NMR analysis (Garson and Staunton 1979) are replacing it.

Nonetheless, if accurate quantitative data on the incorporation of labelled atoms into a metabolite is needed, but precise location of the label is not important, the use of radioisotopes is still indicated. Also it is the method of choice for establishing the most efficient precursor feeding regime prior to ^{13}C labelling studies (see Sect. 5.1).

4 ^{13}C NMR Spectroscopy

The most important stable isotopes in biosynthetic studies are ^{13}C, ^{2}H, ^{18}O and ^{15}N. The advent of high field Fourier-transform spectrometers has allowed the development of ^{13}C NMR spectroscopy to provide a non-destructive detection system, capable of determining the location and relative concentration of each chemically non-equivalent carbon in a molecule, as will be shown. It also enables the presence of ^{2}H, ^{18}O, and ^{15}N to be demonstrated by using isotope-induced shifts in ^{13}C NMR resonance frequencies or by spin–spin coupling to ^{13}C.

A large number of texts (Levy et al. 1980; Abraham and Loftus 1978; Breitmaier and Voelter 1978; Stothers 1972) are available which describe the basic principles of ^{13}C NMR spectroscopy in greater or lesser detail, so only the essential details are summarised here. Since the ^{13}C nucleus has a nuclear spin ($I = ½$), nuclear magnetic resonance signals may be observed, and with the development of sensitive instruments with the means to accumulate (and thus enhance) weak signals, the amount of data available on ^{13}C chemical shifts and spin–spin coupling constants (Marshall 1983) to ^{13}C, ^{1}H, and ^{15}N is increasing rapidly. The undecoupled ^{13}C NMR spectrum shows coupling to both directly bonded hydrogens and longer-range couplings to non-bonded hydrogens. Analysis of these couplings is often of great utility in both structural elucidation and spectral assignments (see Sect. 4.1.3) but the spectra are usually proton noise-decoupled so that a single line is obtained for each carbon in the molecule. Thus, although the natural abundance of ^{13}C is only 1.1%, it is now relatively easy to obtain ^{13}C NMR spectra for any compound given a minimal amount of material (e.g., 50 mg of a compound of molecular weight 500). One great advantage of ^{13}C NMR in the considerable spread of chemical shifts commonly observed: about 200 ppm compared with 10 ppm for ^{1}H shifts. These shifts are of the same relative order as ^{1}H shifts, i.e. alkyl carbons have shifts at higher field (smaller δ value) than aromatic or olefinic carbons, and carbons carrying oxygen or other electronegative substituents have larger δ values than those bearing only carbon or hydrogen substituents. Carbonyl carbons usually provide the lowest field signals. Thus even complex molecules generally give proton noise-decoupled (PND) ^{13}C NMR spectra in which a single line is observed for each carbon in the molecule and co-incident lines are relatively rare.

4.1 Assignment of ^{13}C NMR Spectra

4.1.1 Known Chemical Shifts and Substituent Chemical Shift Effects

The chemical shifts of different functional types fall into well defined ranges; carbonyl carbons resonate at low field (ca. 200 ppm), aromatic and olefinic carbons at 160–100 ppm, aliphatic carbons with electronegative substituents at 90–50 ppm, and simple aliphatics at highest field 50–10 ppm. Substituent effects are generally found to be additive, and rules for predicting chemical shifts in hydrocarbons and aromatics are available. Useful compilations of chemical shifts are available (Breitmaier et al. 1979).

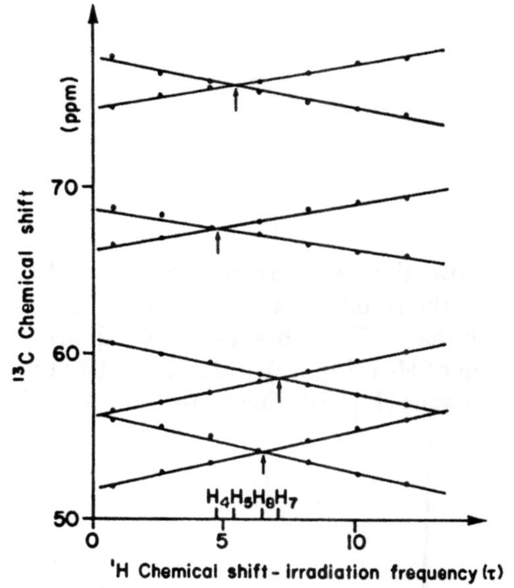

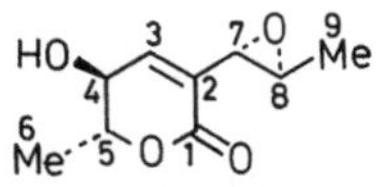

(5)

Fig. 1. Plot of peak frequencies in the ^{1}H off-resonance selectively decoupled ^{13}C NMR spectra of aspyrone **(5)** as a function of ^{1}H irradiation frequency

4.1.2 Off-Resonance and Specific Proton Decoupling

In the PND spectrum all ^{1}H–^{13}C coupling information is lost. In the off-resonance decoupled spectrum, the ^{1}H irradiation is kept at high power levels but the decoupling frequency is moved ca. 500 Hz away from the protons being irradiated, so that one bond ^{1}H–^{13}C coupling patterns partially return and so quaternary, methine, methylene and methyl carbons appear as singlets, doublets, triplets and quartets respectively. If the ^{1}H NMR spectrum has been fully or partially assigned, the ^{1}H and ^{13}C resonances can be interrelated by single-frequency decoupling with the decoupler set at the exact ^{1}H frequency. The attached carbon appears as a singlet in the ^{13}C spectrum, whereas the remaining carbons show off-resonance patterns. This process becomes time-consuming if many resonances require decoupling, but can be overcome by plotting the line frequencies in the ^{13}C NMR spectrum as the ^{1}H irradiating frequency is stepped through the ^{1}H NMR spectrum. Where the lines cross gives the exact ^{1}H and ^{13}C frequencies (Birdsall 1972). The four oxygen-bearing carbons in aspyrone **(5)** have been assigned in this way, as shown in Fig. 1 (Holker and Simpson 1981).

4.1.3 Analysis of Long-Range ^{1}H–^{13}C Couplings

A particularly useful method for both spectral assignment and structural elucidation is to analyse the origins of long-range, i.e., greater than one-bond ^{1}H–^{13}C couplings using specific, low-power decoupling to selectively remove long-range couplings to ^{13}C only from the hydrogen which is being irradiated. This has been used to great effect in the structural reassignment and spectral assignment of the antibiotic LL-D253α **(6)**, a chromanone metabolite of *Phoma pigmentivora*

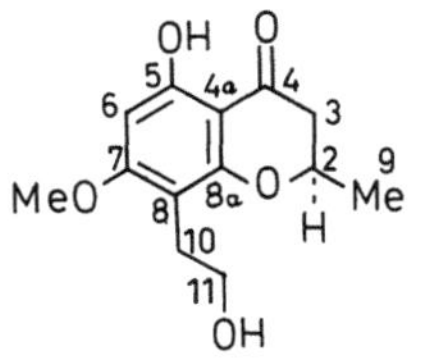

(6)

(McIntyre and Simpson 1984). Figure 2 shows the downfield region of the fully ^{1}H coupled ^{13}C NMR spectrum of **(6)** and the results of a series of specific decoupling experiments. Thus it may be seen that C-7 which appears as a quartet in Fig. 2a simplifies to triplet on irradiation of H-6 (Fig. 2e) and to a doublet on irradiation of the benzylic methylene hydrogens (Fig. 2d) due to the removal of

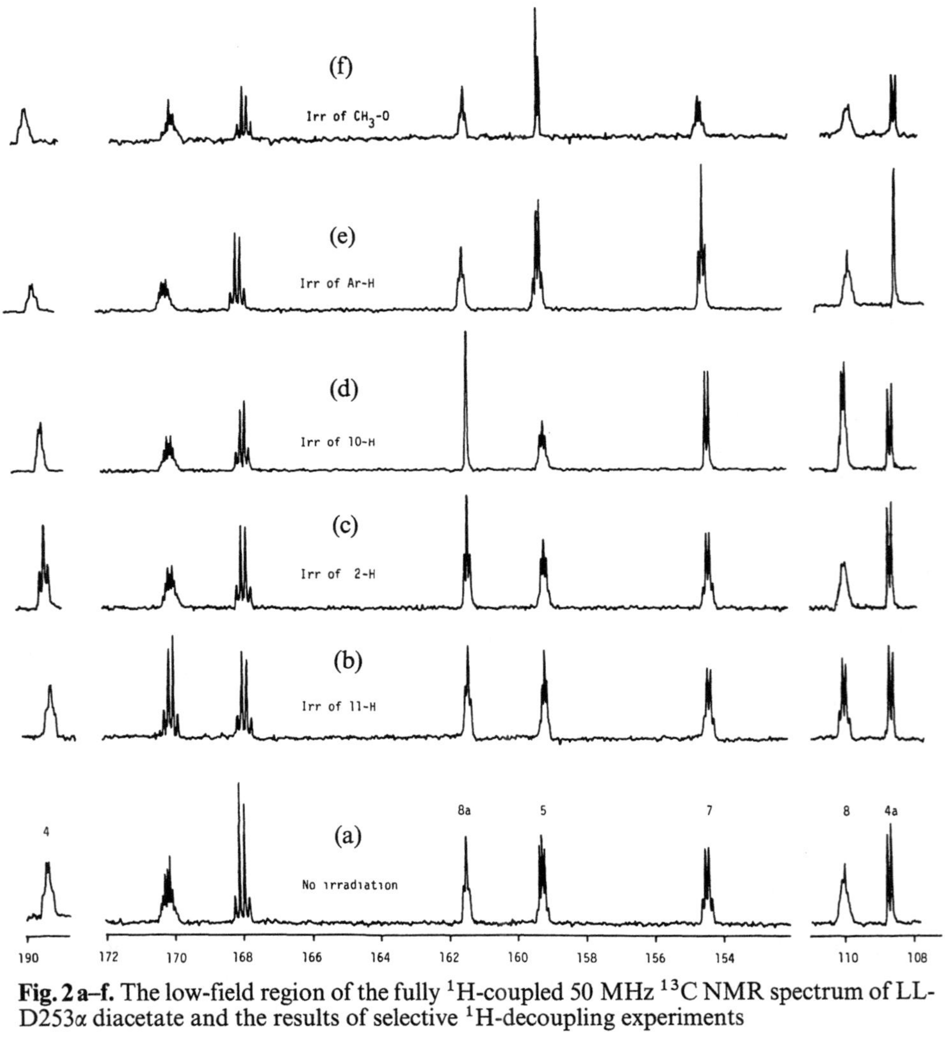

Fig. 2 a–f. The low-field region of the fully ^{1}H-coupled 50 MHz ^{13}C NMR spectrum of LL-D253α diacetate and the results of selective ^{1}H-decoupling experiments

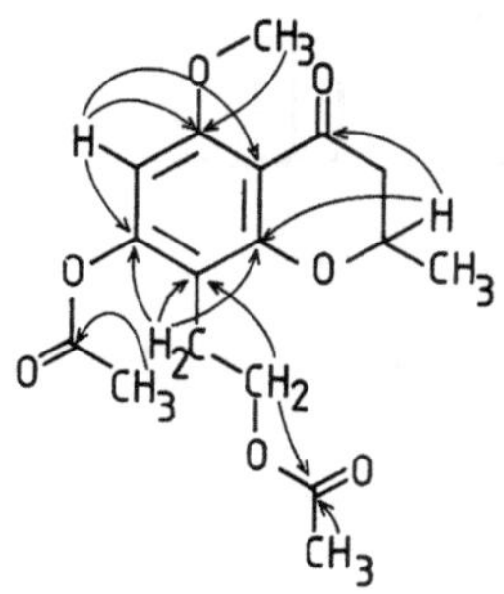

Fig. 3. Long-range ^{1}H-^{13}C couplings in LL-D253α diacetate identified by selective ^{1}H-decoupling experiments (Fig. 2)

2- and 3-bond couplings respectively. Irradiation of H-6 simplifies the pentet due to C-5 in Fig. 2a to a quartet (Fig. 2e). This quartet splitting is due to coupling to the methoxyl hydrogens as an irradiation of these the signal due to C-5 now appears as a doublet (Fig. 2f). Similarly, it may be seen that C-8 which appears as a multiplet in Fig. 2a shows 2-bond coupling to the C-10 methylene hydrogens and 3-bond couplings to H-6 and to the C-11 methylene hydrogens. The long-range couplings which have been identified in this way are summarised in Fig. 3. These permit only one structure and simultaneously assign the ^{13}C NMR spectrum prior to biosynthetic studies.

4.1.4 Polarisation Transfer and 2D NMR Methods (Benn and Gunther 1983)

The methods described in Sections 4.1.2 and 4.1.3 have been partly superceded, e.g. by polarisation transfer methods such as DEPT and INEPT which provide a more sensitive method of identifying and distinguishing methyl, methylene, methine and quaternary carbons. These methods utilise pulse sequences which enable only those carbons with a given number of hydrogens attached to be observed selectively. Thus they can be used to simplify congested spectral regions so that only carbons bearing one, two or three, or indeed no hydrogens, are observed. This is illustrated in Fig. 4 which shows the DEPT spectra of austin **(7)**, a mycotoxin isolated from *Aspergillus ustus*. Correlation of ^{1}H and ^{13}C chemical shifts can be achieved by two-dimensional heteroscalar correlation NMR spectra. Figure 5 shows how the ^{1}H and ^{13}C resonances for the methyl groups in terretonin **(8)**, a metabolite of *Aspergillus terreus*, can be matched. Other pulse sequences such as 2D INADEQUATE permit carbon connectivity patterns to be established. These methods have been very well reviewed (Benn and Gunther 1983) but

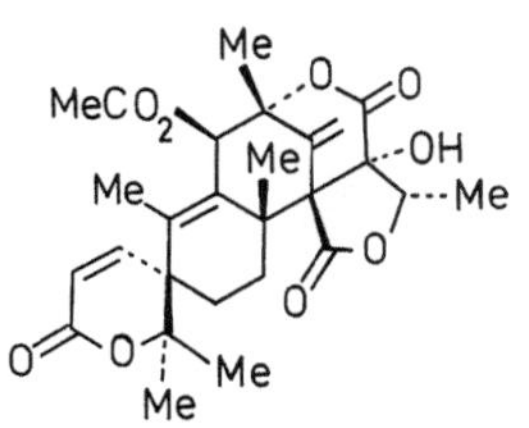

(7)

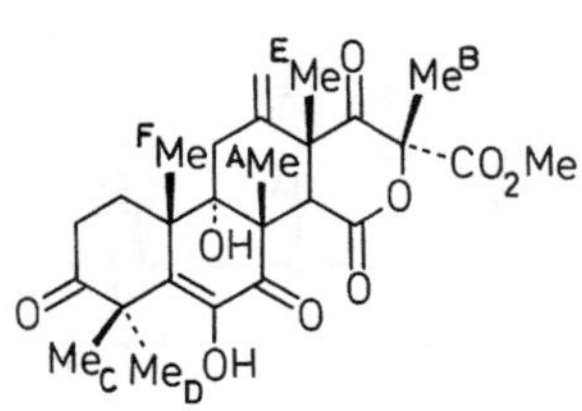

(8)

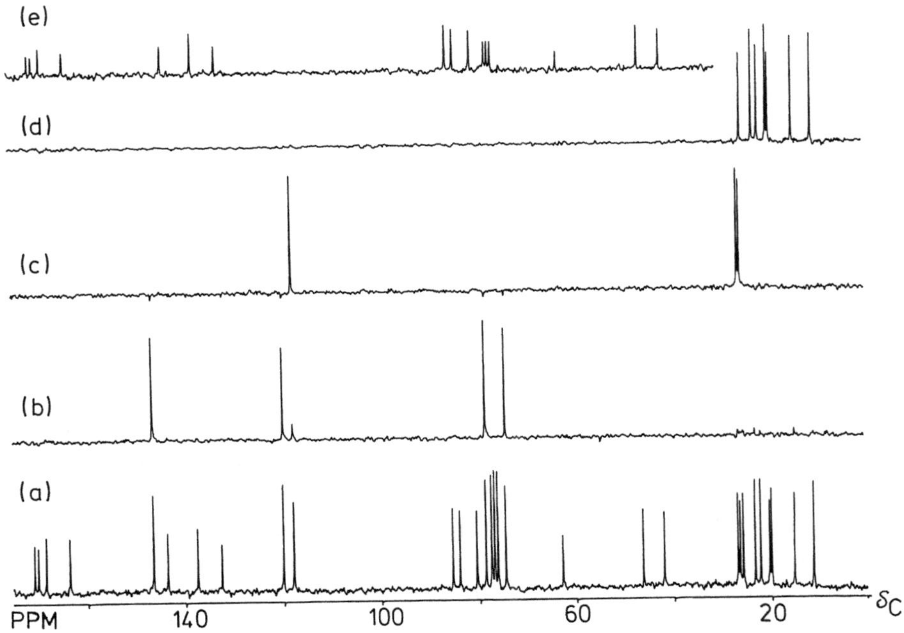

Fig. 4. **a** Normal PND ^{13}C NMR spectrum of austin **(7)** and DEPT spectra showing **b** methines only, **c** methylenes only, **d** methyls only and **e** quaternary carbons only

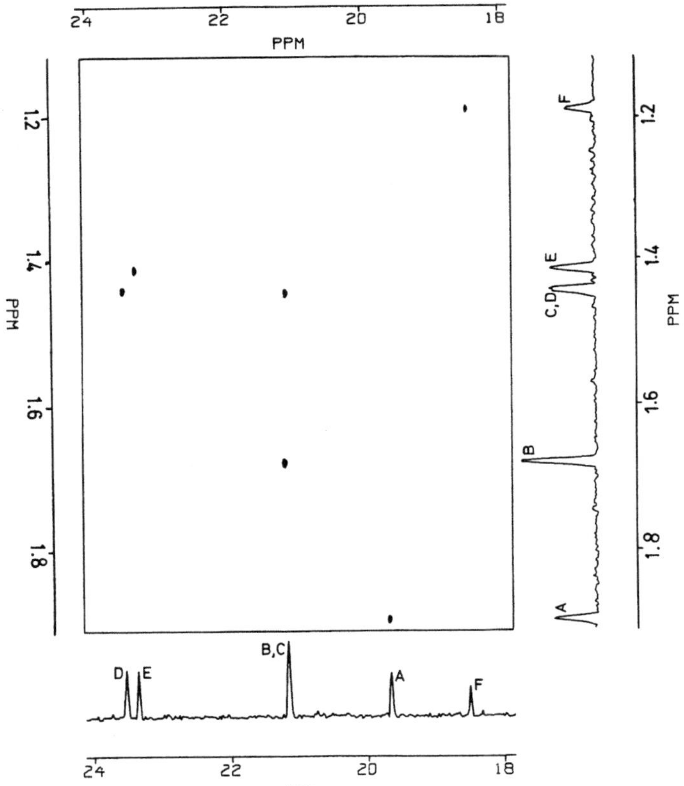

Fig. 5. Two-dimensional heteroscalar correlation NMR spectrum of terretonin **(8)** matching the ^1H and ^{13}C resonances of each methyl group

it should be noted that they require not only sophisticated spectrometers with very good data-handling facilities, but also highly skilled operators. They are far from being the routine methods that the literature often implies.

4.1.5 Lanthanide-Induced Shift Studies

These are not as useful in ^{13}C studies as in ^{1}H NMR because, as for solvent and anisotropy effects, the actual shifts are of the same absolute value in both and so are relatively small in ^{13}C NMR. However, they can be useful for separating overlapping resonances. In complicatic acid **(9)** the resonances due to C-3 and C-10

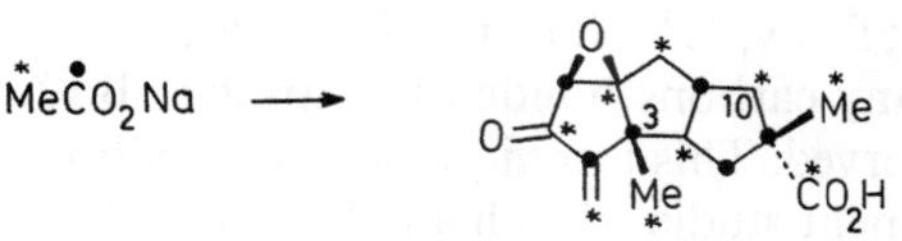

(9)

both occur at 46 ppm. However addition of Eu(fod)$_3$ separated these resonances and showed that C-3 but not C-10 was enriched from [1-^{13}C]acetate in cultures of *Stereum complicatum* (Feline et al. 1974).

Another important use of shift reagents is to resolve the ^{1}H NMR spectrum prior to specific proton decoupling studies on the ^{13}C NMR spectrum.

4.1.6 Model and Derivative Studies

Model compounds whose chemical are known can be helpful in assigning the spectrum of a new compound, though they must be used with care. Quaternary carbons always present the greatest difficulties in assignment, and studying the variation in chemical shifts in a series of closely related compounds may be the only method of reaching an assignment. The ^{13}C NMR spectrum of tajixanthone

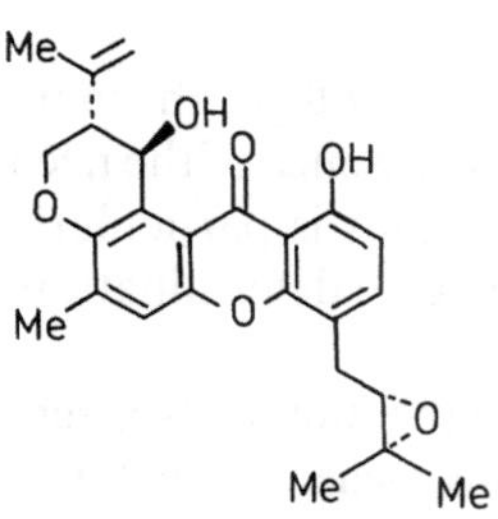

(10)

(10), a prenylated xanthone, was fully assigned by a study of eleven derivatives prior to biosynthetic studies (Holker et al. 1974).

4.1.7 Synthesis of Isotopically Labelled Compounds

Synthesis of a compound enriched at a known site with ^{13}C or ^{2}H may be used for assignment. This has the effect either of enhancing the signal of the labelled carbon or deleting the signal from the spectrum in the case of ^{2}H substitution.

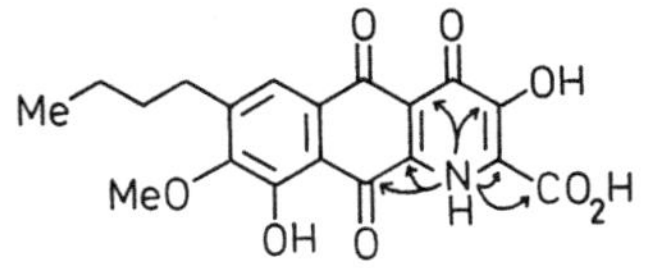

(11)

Fig. 6. ^{13}C–^{15}N couplings observed in the PND ^{13}C NMR spectrum of ^{15}N-enriched phomazarin

4.1.8 Incorporation Studies

^{15}N is an isotope of spin ½ and so is NMR active and will couple with ^{13}C nuclei. Thus a metabolite grown in the presence of, say, $K^{15}NO_3$ may incorporate ^{15}N and will exhibit ^{15}N–^{13}C couplings for any carbons bonded to nitrogen. Both one- and two-bond couplings may be observed. This technique was used in both structural elucidation and spectral assignment studies (Birch and Simpson 1979) of phomazarin **(11)**. The couplings observed are indicated in Fig. 6. Obviously the ^{15}N may also be incorporated by chemical synthesis.

Similarly any compound with adjacent ^{13}C nuclei will exhibit ^{13}C–^{13}C couplings (see Sect. 6.2.2) between these nuclei. Again, the ^{13}C may be incorporated metabolically or by synthesis.

5 Biosynthetic Methodology

The availability of compounds enriched with ^{13}C and other stable isotopes has increased rapidly and a wide range is now available at enrichments of up to 99%. In addition, many others may be readily synthesised by standard methods (Ott 1980) from simple precursors such as $^{13}CO_2$, $^{13}CH_3I$, $K^{13}CN$, $^{15}NH_4Cl$, and $H_2{}^{18}O$.

Quite apart from the convenience of using a non-radioactive isotope, the main advantage of ^{13}C NMR studies over ^{14}C radio-labelling is the absence of the need for chemical degradations to establish the sites of labelling. This is particularly important as degradative chemistry is now little used in structure determination, whereas ^{13}C NMR is routinely used for this purpose.

A disadvantage of ^{13}C NMR, one not shared by radio-labelling, is its insensitivity. This is due mostly to the low energy of the transitions measured in NMR generally, but the problem is more acute in the case of ^{13}C than for protons because of the smaller magnetogyric ratio (γ) of ^{13}C (only 0.252 times that of proton). Sensitivity, atom for atom, is proportional to γ^3. In addition, the natural abundance of ^{13}C is only 1.1%. This is a mixed blessing. It means that natural abundance spectra are reasonably easy to obtain without the complications of ^{13}C–^{13}C coupling; however, it also requires that very efficient incorporation of precursors into products is achieved. For this reason, the method has found more limited application in plant studies, as high incorporations are difficult to achieve and whatever labelled precursor is incorporated tends to become diluted by large pools of endogenous material.

5.1 Precursor Incorporation (Simpson 1975)

Before carrying out experiments with stable isotopes, it is necessary to carry out extensive experiments to optimise conditions for feeding of labelled substrates. It is routine procedure to perform these preliminary experiments with ^{14}C-labelled precursors. Precursor efficiency may be assessed in several ways in biosynthetic experiments (Brown 1972) but for ^{13}C studies the important criterion is *dilution* of added label. For ^{14}C, dilution per labelled site is given by:

$$\frac{\text{specific activity of precursor}}{\text{specific activity of product}} \times \text{no. of labelled sites}$$

To obtain unequivocal results in ^{13}C studies using typically 90% enriched precursors, dilutions per labelled site of ca. 100 or less are therefore required (see Sect. 6.1.1).

Three main parameters require studying: time of precursor addition, period of growth after addition, and amount of precursor required. Maximum incorporation usually occurs when precursors are added at the start of maximum metabolite production, e.g. the start of the idiophase in the case of microbial fermentations. This necessitates determination of cell growth and metabolite production curves. The period of growth after addition of precursor may be critical. If the turnover of exogenous precursor is rapid, then prolonged growth results merely in increased dilution of labelled metabolite by further metabolite production from unlabelled endogenous substrate. Thus it is important to note that neither maximum yield of metabolite or maximum total incorporation of label is the important factor, the prime consideration being to obtain the *minimum dilution* of label given a *sufficient yield* of metabolite for the ^{13}C NMR spectrum to be determined. In general, dilution decreases with increasing amounts of precursor. Thus mass versus incorporation studies will determine the minimum amount of labelled precursor that must be added to obtain a satisfactory enrichment.

The amount of labelled precursor that can be used is limited both by the actual expense of the isotopic label and also due to the adverse effects that high levels of added precursor may have on the metabolism of the organism and may in fact change the biosynthetic pathway under study. Toxicity effects may be overcome by pulsed addition of precursor or by continuous slow addition over a prolonged period by use of e.g. peristaltic pumps.

6 ^{13}C Enrichment Studies

The labelling pattern resulting from incorporation of ^{13}C-enriched precursors is determined by obtaining the PND ^{13}C NMR spectrum of the labelled metabolite and comparing it with the spectrum of the unlabelled metabolite. A large number of different types of experiments are now available.

6.1 Single ^{13}C-Labelling

This was the first method to be developed and remains the simplest. It is best understood by examining a formal model system.

If we consider any four contiguous carbons in a polyketide derived molecule, normally these carbons will be derived from endogenous acetyl CoA produced by

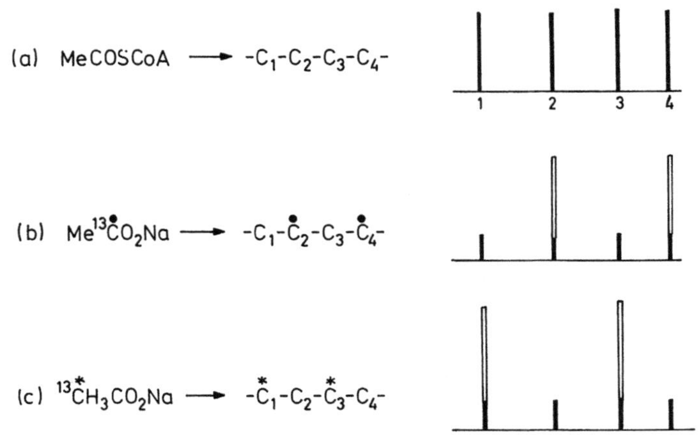

Fig. 7 a–c. Simulated PND ^{13}C NMR spectra of a polyketide-derived moiety **a** at natural abundance, **b** enriched from [1-^{13}C]acetate, and **c** enriched from [2-^{13}C]acetate

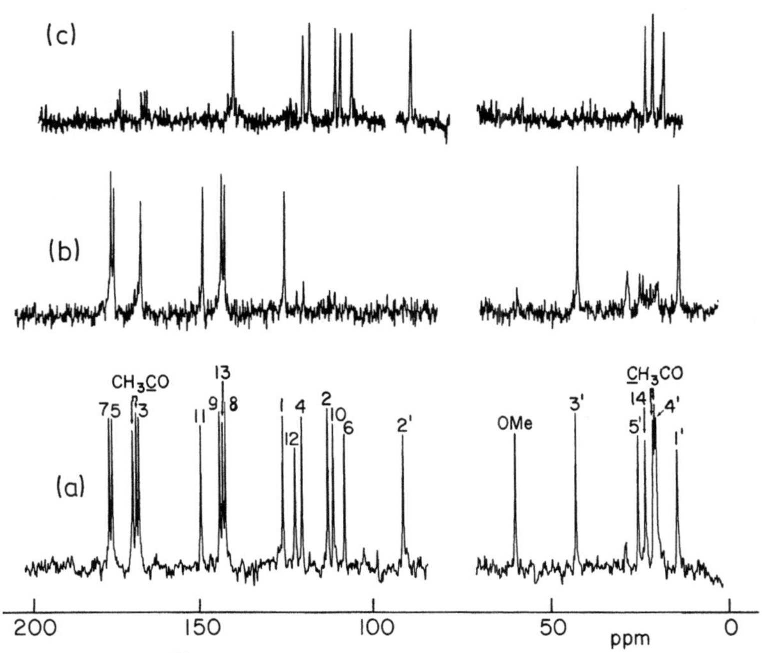

Fig. 8 a–c. PND ^{13}C NMR spectra of the diacetate of deoxyherqueirone: **a** at natural abundance, **b** enriched from [1-^{13}C]acetate, and **c** enriched from [2-^{13}C]acetate

the cell's normal metabolism and will contain only natural abundance ^{13}C (1.1%) and so in the PND ^{13}C NMR spectrum each carbon will give rise to one sharp line of more or less equal intensity, Fig. 7a. Now, if sodium acetate, in which the carboxyl carbon is highly enriched (ca. 95%) with ^{13}C, ([1-^{13}C]acetate) is added, then this exogenous acetate will be diluted to a greater or lesser extent by the endogenous acetate pool, but some of it will be incorporated into the metabolite and so those carbons, say C-2 and C-4, which were originally derived from the carboxyl carbon of acetate, will contain extra ^{13}C and this will manifest itself as an increase in the appropriate signal intensities in the ^{13}C NMR spectrum of the enriched metabolite, Fig. 7b. If acetate in which the methyl carbon is enriched ([2-^{13}C]acetate) is now added, then enhancement of the alternate signals is obtained, Fig. 7c. Thus, simply by feeding ^{13}C-labelled precursor and determining the ^{13}C NMR spectrum, the signals which show enhanced intensities indicate the sites of enrichment provided the ^{13}C NMR spectrum has been assigned unambiguously. It is important to note that this does not give us any further information than (in theory at least) could have been obtained by classical ^{14}C techniques, though in practice it is easier and usually more comprehensive.

An example of single labelling studies with acetate is the incorporation of [1-^{13}C]- and [2-^{13}C]acetates into deoxyherqueinone **(12)**, a metabolite of *Penicillium*

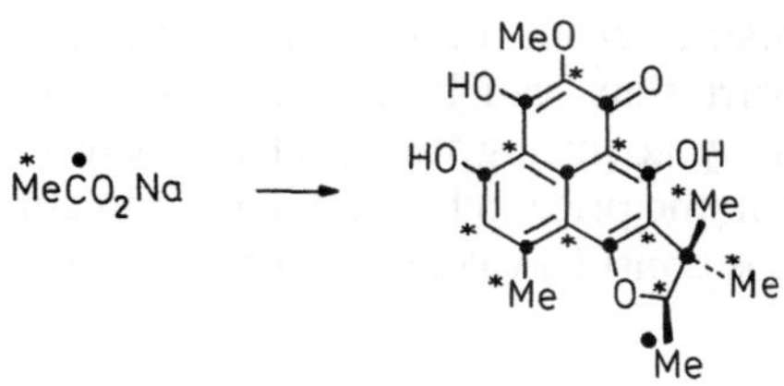

(12)

herquei (Simpson 1979). The resultant spectra are shown in Fig. 8. If high enrichments are obtained, as in this case, the labelled sites are readily apparent from visual inspection of the spectrum. However, in general, enrichments are lower and identification of enriched sites with certainty can be more difficult, as discussed below.

6.1.1 Quantitative ^{13}C Measurements

In principle at least, quantitative values for the degree of enrichment by labelled precursors and intermediates can be determined by comparing the integrated intensities of different carbon resonances in the PND ^{13}C NMR spectrum. The percentage of ^{13}C isotope above natural abundance has been defined as:

$$\text{\%age enrichment} = 1.1 \times \frac{\text{(integrated intensity at labelled centre)}}{\text{(integrated intensity at unlabelled centre)}} - 1.1\,.$$

In practice, however, comparing signal heights or integrals is problematic and the complications have been widely discussed (Abraham and Loftus 1978; Simpson 1975). Due to the wide range of spin-lattice relaxation times (T_1's) encountered in ^{13}C NMR spectroscopy and the variable effects of the nuclear Overhauser enhancement (NOE) induced by proton decoupling, the line intensities in a ^{13}C spectrum usually show wide variation even in a natural abundance spectrum. Fourier transform NMR spectrometers exacerbate this problem using normal pulse sequences and add another one: the digitization of data often results in a poorly defined peak shape. The first problem can be circumvented to an extent by comparing unenriched and enriched resonances of carbons in a similar environment, but a better solution is to obtain natural abundance and enriched spectra under identical conditions, normalise both spectra to a standard which is known to be unlabelled, and compare the intensities of the lines in the two spectra directly. An alternative approach is to add a paramagnetic relaxation agent such as chromium Tris-acetoacetonate which suppresses the NOE and shortens the T_1's. It may also be necessary to use "inverse gated" proton decoupling in which the NOE is reduced by only decoupling during acquisition. Further alleviation of the effects of long T_1's may be achieved by the use of a long delay between pulses.

The second problem, that of data handling, manifests itself in random variations in intensities in the lines of spectra obtained from the same sample under the same experimental parameters. These variations can be reduced by using more data points to define the spectrum or by acquiring the spectrum several times and averaging the results. As can be seen, the price to be paid in instrument time, in return for reliable enrichment data, is considerable, and some compromise is generally required.

6.1.2 Single ^{13}C-Labelling Studies in Plants

Due to the reasons discussed above, ^{13}C-labelling studies are much more difficult to carry out on higher plant metabolites than on microbial metabolites. Some of the problems associated with precursor incorporation and transport, and dilution by endogenous pools of precursors and metabolites can be overcome by use of cell-free enzyme systems and in particular by use of tissue cultures. Obviously both of these require much effort to develop workable systems. However, there have been a number of successful applications, particularly in the alkaloid field and in the important area of phytoalexin metabolites.

^{14}C-Labelling studies indicated that the lactam **(14)** was an intermediate in the biosynthesis of camptothecin **(15)** in *Camptotheca acuminata*. Owing to the ab-

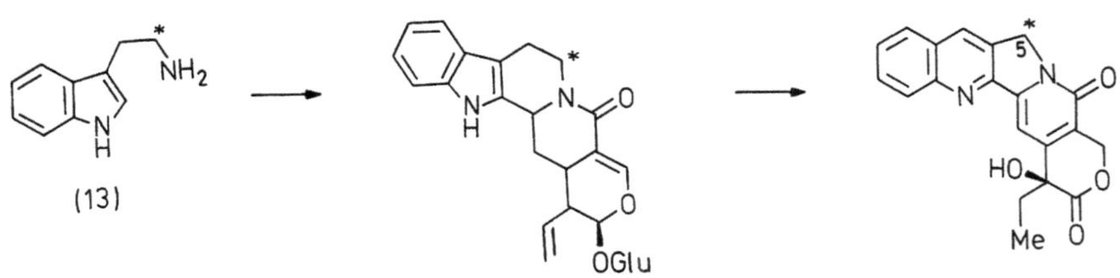

(13) (14) (15)

sence of suitable degradations, ^{13}C NMR was used to prove specificity of incorporation. Thus, [1-^{13}C]tryptamine **(13)** was synthesised from $K^{13}CN$ and converted to the [5-^{13}C]lactam **(14)**, 38 mg of which was wick-fed to intact plants. After 2 days growth, 20 mg of comptothecin was isolated which showed a ca. 55% enhancement of the C-5 resonance only (Hutchinson et al. 1974).

[1-^{13}C]Autamnaline **(16)** was synthesised and injected as the hydrochloride (300 mg) into seed capsules (1 mg per capsule) of *Colchicum autumnale*. After 2

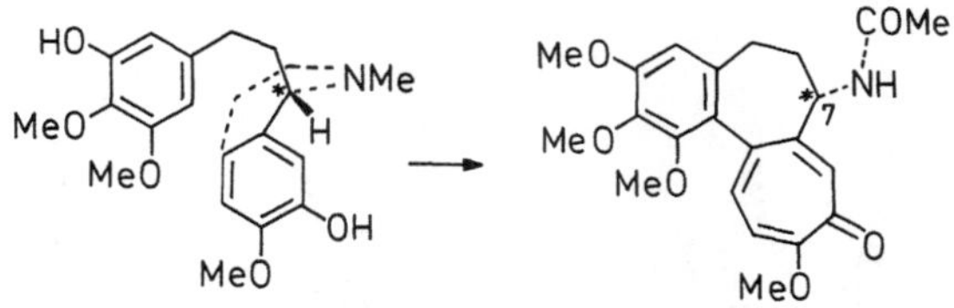

(16) (17)

(18) (19)

weeks' growth 1.24 g colchicine **(17)** was isolated. The resultant ^{13}C NMR spectrum showed a 2.5-fold enhancement of the C-7 signal only (Battersby et al. 1974).

By feeding [2′-^{13}C,^{14}C]anatabine **(18)** to various *Nicotania* species it was established that this alkaloid is a precursor of the $\alpha\beta$-bipyridyl **(19)**, the ^{13}C label being detected by the enrichment observed for C-2 (10%) by ^{13}C NMR. The precursor was fed to plants in hydroponics. However the transformation was only observed in plants which had been allowed to dry in the sun for several days and no conversion could be detected in freshly harvested plants so it is not clear whether it is an enzymic process or simply due to aerial oxidation (Leete et al. 1979).

Feeding [N-Methyl-^{13}C]stylopine α-methochloride **(20)** to *Chelidonium majus* resulted in a 10% enrichment of the N-methyl signal of protopine **(21)** and a 29%

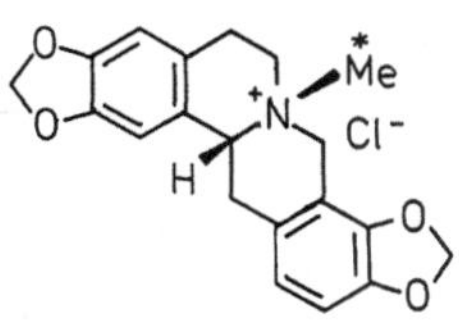

(20)

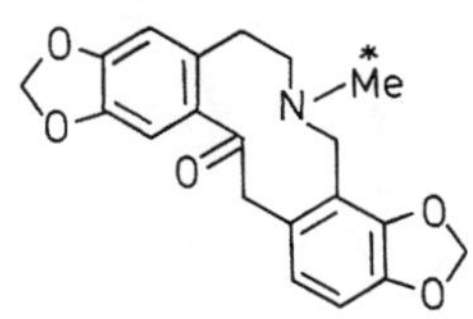

(21)

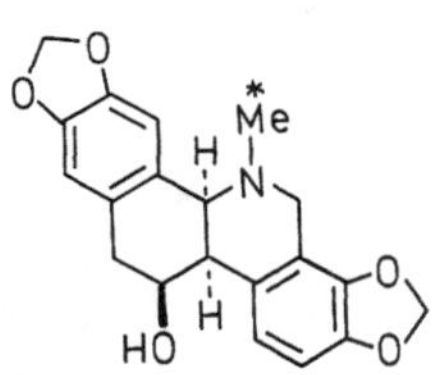

(22)

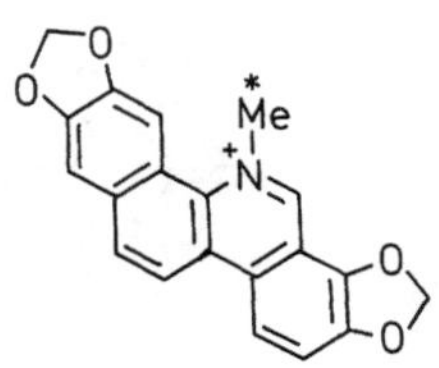

(23)

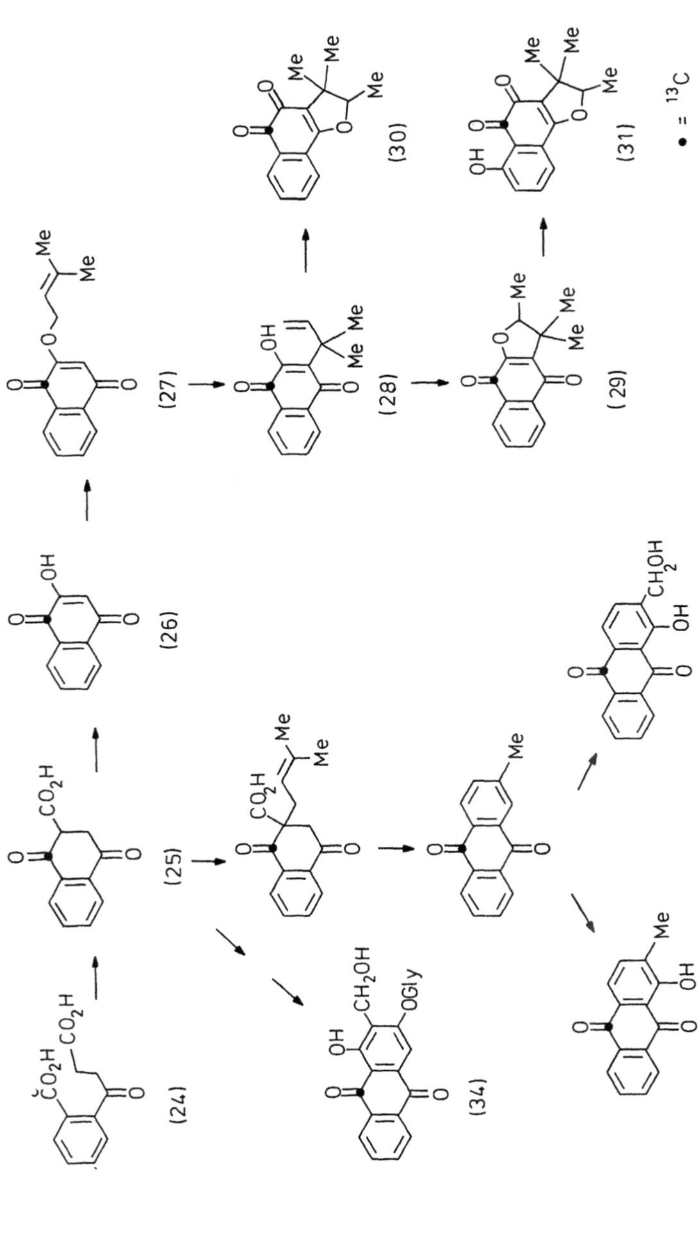

Scheme 3

Scheme 4

enrichment of the N-methyl signal of chelidonine **(22)**. Similarly [N-methyl-^{13}C]protopine enriched the N-methyl signals of chelidonine **(22)** and sanguinarene **(23)** by ca. 40% in each case. Other transformations are reported in this paper but the level of enrichments observed are probably too low to be regarded as certain (Takao et al. 1976).

The isolation of a range of naphthoquinones and anthraquinones from plant cell cultures has allowed their biosynthesis to be studied by ^{13}C NMR. Thus, o-succinylbenzoic acid **(24)**, labelled as shown in was incorporated into dunnione **(30)**, α-duninone **(29)** and 8-hydroxydunnione **(31)**, and the anthraquinones **(32)** and **(33)** as shown. The location of the introduced label led to the conclusion that the naphthoquinones were biosynthesised via **(25)** and lawsone **(26)** which is O-prenylated to give **(27)** (Scheme 3). A Claisen-type rearrangement then leads to **(28)** which can then be converted into **(29)** or into **(30)** and **(31)** (Inone et al. 1982). The anthraquinones appear to be produced by prenylation at C-2 of **(25)** as shown. However, feeding ^{13}C-labelled o-succinylbenzoic acid to cell cultures of *Gallium mollugo* gave lucidin primeveroside **(34)** labelled as shown demonstrating that prenylation occurs at C-3 in this case (Inoye et al. 1979).

3,10-Dihydro-1,4-dimethylazulene **(35)** is an unusual product of cultured cells of the liverwort *Calyprogeia granulata* (Takeda and Katoh 1983). Biosynthetic studies using [2-^{13}C]acetate have established a terpenoid origin for **(35)** and its co-metabolite **(36)**.

6.2 ^{13}C–^{13}C Spin-Spin Coupling

In natural abundance ^{13}C NMR spectra ^{13}C–^{13}C spin-spin coupling is not normally observed as the probability of ^{13}C nuclei being adjacent is equal to the square of the natural abundance, giving satellites of 0.55% of the intensity of the main signal. They can be observed by resorting to the double quantum resonance technique, INADEQUATE, which allow observation *only* of coupled signals (Benn and Gunther 1983), but this requires very large samples and sophisticated spectrometers. However, in enriched metabolites, the probability of having adjacent ^{13}C nuclei is much higher and the detection of a ^{13}C–^{13}C coupling can provide conclusive evidence that two labels have been incorporated into adjacent positions in a molecule. This is normally done by using doubly labelled precursors, but ^{13}C–^{13}C couplings can also arise from administration of singly labelled precursors.

6.2.1 Singly ^{13}C-Labelled Precursors

Molecular rearrangement of a biosynthetic intermediate may give rise to a ^{13}C–^{13}C coupling. Aspyrone **(5)**, a metabolite of *Aspergillus melleus* (Holker and Simpson 1981), when enriched from [2-^{13}C]acetate shows a coupling of 61 Hz between C-2 and C-7 (Fig. 9b). This coupling arises due to an intramolecular rearrangement of the precursor pentaketide chain as shown in Scheme 4.

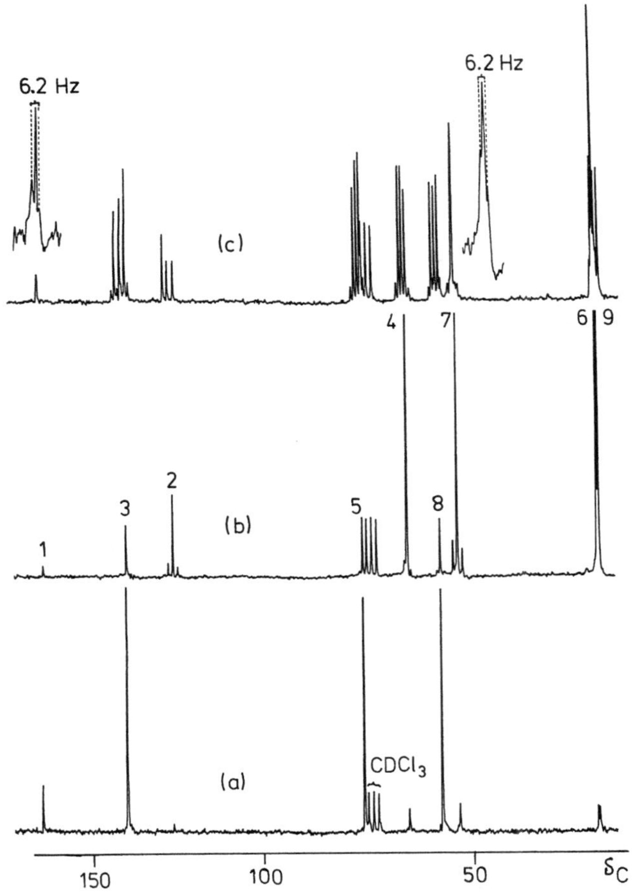

Fig. 9 a–c. PND ^{13}C NMR spectra of aspyrone **(5)** enriched from: **a** [1-^{13}C]acetate, **b** [2-^{13}C]acetate, and **c** [1,2-$^{13}C_2$]acetate

Incorporation of [1-^{13}C]acetate into the sesquiterpene dihydrobotrydial **(38)** in *Botrytis cinerea* resulted in a coupling being observed between C-7 and C-8 (Bradshaw et al. 1977). Conversion of [2-^{13}C]acetate into succinate in the Krebs cycle results in a ^{13}C–^{13}C coupling between C-11 and C-15 in avenaciolide **(39)** (Tanabe et al. 1973). Similar metabolic transformations give rise to ^{13}C–^{13}C cou-

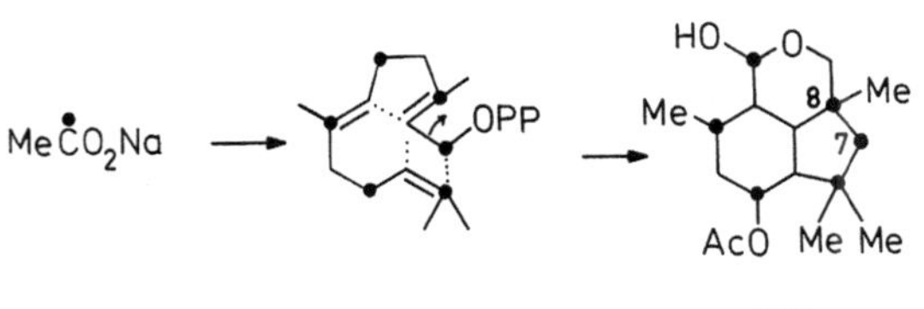

(38)

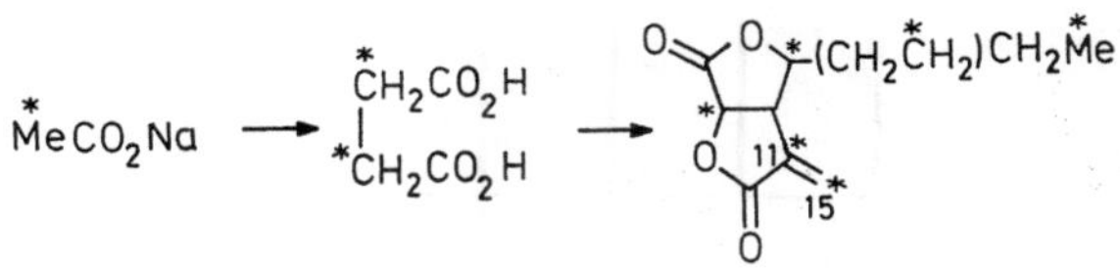

(39)

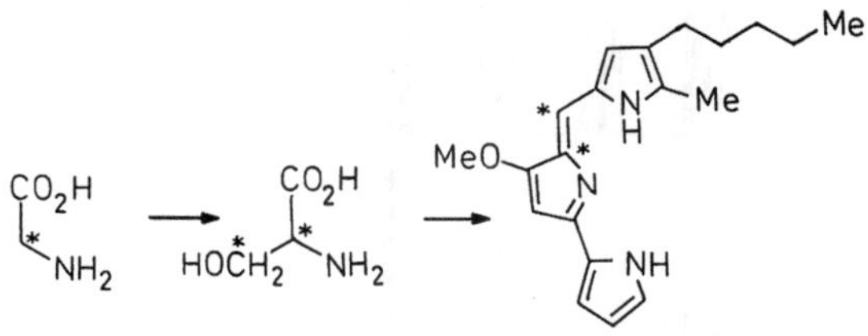

(40)

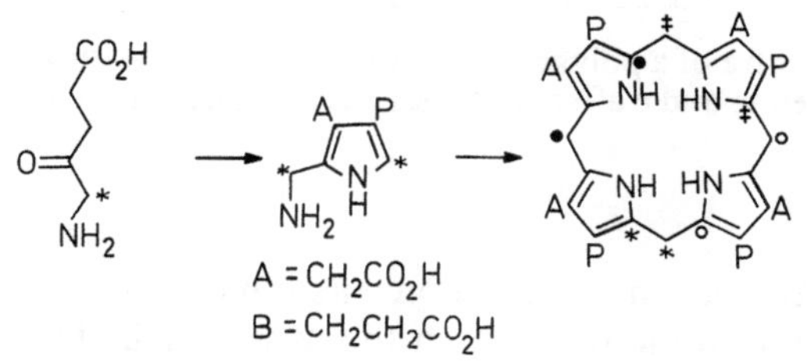

(41) (42) (43)

plings when [2-^{13}C]glycine is incorporated via serine into prodigiosin **(40)** (Wasserman 1974) and [5-^{13}C]-δ-aminolaevulinic acid **(41)** is incorporated into vitamin B_{12} **(43)** via porphobilinogen **(42)** (Brown et al. 1972). However, it is important to note that all of these were in effect fortuitous observations, resulting from high specific incorporation of labelled precursors rather than by experimental design.

6.2.2 Double ^{13}C-Labelled Precursors

This has been one of the major developments in biosynthetic methodology and permits information to be obtained which would have been impossible or at best extremely difficult to obtain by classical radio-isotope labelling techniques. Again, the basic concept can be illustrated by consideration of a model polyketide system (Fig. 10).

If we consider a molecule of acetate in which both carbons are entirely ^{13}C, ([1,2-$^{13}C_2$]acetate), it contains two adjacent nuclei of spin ½ and so they will couple to each other. If this acetate molecule is incorporated intact into a metabolite, then in any individual molecule, those pairs of carbons derived from an originally intact acetate unit must necessarily both be enriched simultaneously and so will show a mutual ^{13}C–^{13}C coupling. Thus if C-1 is enriched, then C-2 must also be enriched. In the resultant ^{13}C NMR spectrum, the natural abundance signal is flanked by ^{13}C–^{13}C coupling satellites (Fig. 10b). By analysing the coupling

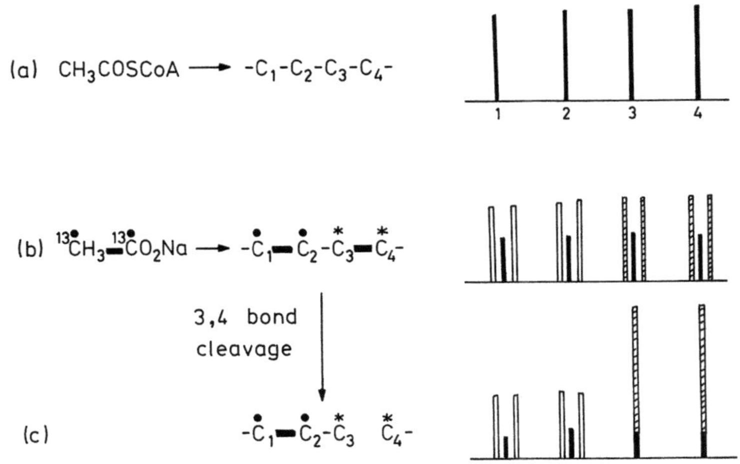

Fig. 10 a–c. Simulated PND ^{13}C NMR spectra of a polyketide derived moiety: **a** at natural abundance, **b** enriched from [1,2-$^{13}C_2$]acetate and **c** after cleavage or rearrangement of an originally intact acetate unit

patterns, information is obtained on the way in which the precursor molecules are assembled on the enzyme surface, and on the way the, in this case, polyketide, chain folds up prior to condensation and cyclisation. If at any stage in the biosynthesis the bond between two carbons originally derived from an intact acetate unit is broken, then the ^{13}C–^{13}C coupling is lost and these carbons then appear simply as enhanced singlets, as shown for C-3 and C-4 in Fig. 10c. In this way bond cleavage and rearrangement processes occurring during biosynthesis can be detected.

^{13}C–^{13}C Couplings are generally between 30 and 80 Hz and increase in proportion to the amount of "s" character of the atoms in the bond. Hence sp^3–sp^3 bonds are typically 35 Hz; sp^2–sp^3 are 45 Hz, and sp^2–sp^2 are 60 Hz (Marshall 1983). Substitution of one or both atoms by oxygen increases the size of the coupling. Usually the couplings will be sufficiently different in magnitude to enable the pairs of coupled carbons to be matched up. In addition, if the spectrum has been unambiguously assigned it will be apparent which carbons are mutually coupled. Couplings may also be confirmed by selective homonuclear ^{13}C–^{13}C decoupling or by a 2D INADEQUATE spectrum (Benn and Gunther 1983).

The contiguous double labelling technique has been used extensively. This is partly because it extends the permissable dilution factor to at least 2000, as small ^{13}C–^{13}C coupling satellites can be observed with more certainty than the corresponding enrichment from a singly labelled precursor. However, its success is largely because of the extra information obtainable in respect of bond cleavages, rearrangements and symmetrical elements of intermediates.

The ^{13}C NMR spectrum resulting from incorporation of [1,2-$^{13}C_2$]acetate into aspyrone **(5)** is shown in Fig. **9c**. ^{13}C–^{13}C couplings of 68, 41 and 44 Hz are observed between C-2 and C-3, C-4 and C-5, and C-8 and C-9 respectively, indicating their derivation from originally intact acetate units. C-1, C-6 and C-7 however appear as enriched singlets indicating their derivation from cleaved acetate units as indicated above in Scheme 4.

Due to its extra sensitivity the technique has been applied to many higher plant metabolites. Some of these studies are discussed briefly below to indicate the type of information these studies can provide.

[1,2-$^{13}C_2$]Acetate was incorporated into the flavone apigenin **(45)** and the flavonol **(46)** by cell suspension cultures of parsley, *Pteroselinum hortense*, with

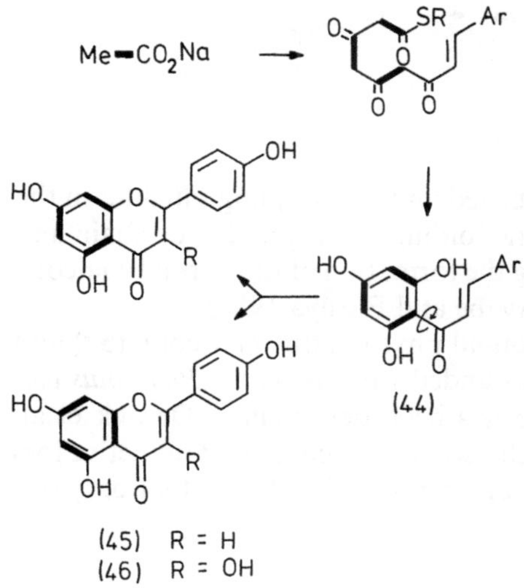

Scheme 5

randomisation of ^{13}C–^{13}C couplings in ring A (Light and Hahlbrock 1980). Each carbon shows two sets of coupling satellites due to coupling to *both* adjacent carbons. This means that a symmetrical intermediate, presumably the chalcone **(44)** in which ring A is free to rotate, is an intermediate in the biosynthesis of **(45)** and **(46)** as shown in Scheme 5. Detection of symmetrical intermediates by this type of randomisation of coupling is one of the more important applications of double ^{13}C-labelling experiments. In contrast, [1,2-$^{13}C_2$]acetate is incorporated into ring A of the phytoalexin pisatin **(47)** in *Pisum satinum* without randomisation, show-

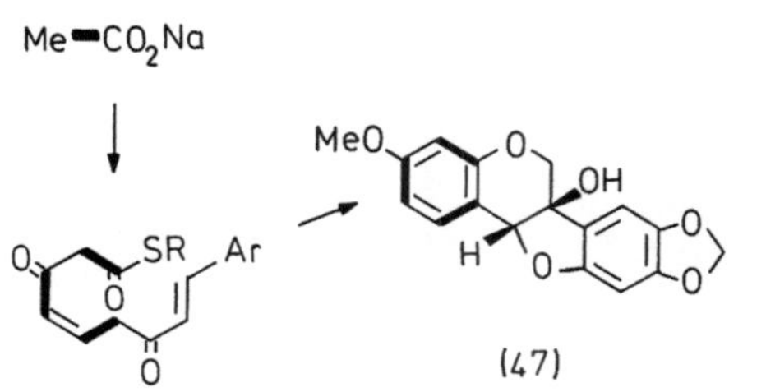

Scheme 6

ing that deoxygenation of the polyketide precursor must occur before cyclisation and aromatisation (Scheme 6) (Stoessl and Stothers 1979). In a similar experiment, the labelling pattern in a sample of the retrochalcone echinatin **(49)** produced when cell cultures of *Glycyrrhiza echinata* were fed [1,2-$^{13}C_2$]acetate was

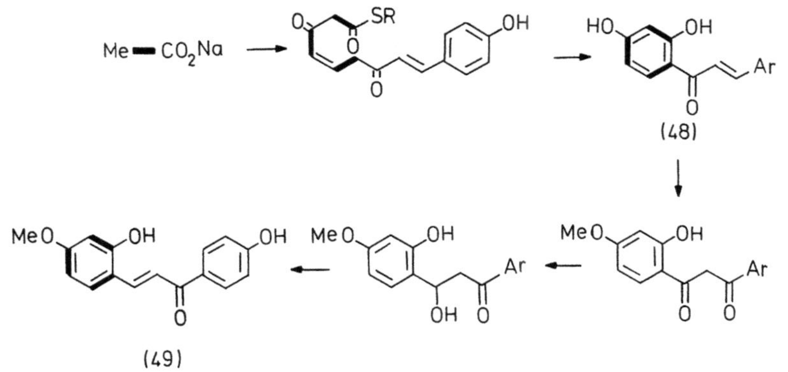

Scheme 7

analysed. Ring B rather than ring A was labelled and the coupling satellites in the ^{13}C NMR spectrum demonstrated a specific folding of the polyketide chain and reductive loss of the oxygen function during the formation of the normal chalcone intermediate **(48)** as shown in Scheme 7 (Ayabe and Furuya 1982).

Patterns of ^{13}C labelling in the isoflavonoid phytoalexins phaseolin **(50)** and kevitone **(51)** produced by cotyledons of wounded French beans (*Phaseolus vulgaris*) in the presence of [1,2-$^{13}C_2$]acetate has been determined (Dewick et al. 1982). These show that specific folding of the polyketide chain and reductive loss of an oxygen atom occur prior to ring closure in phaseolin **(50)** so that only one

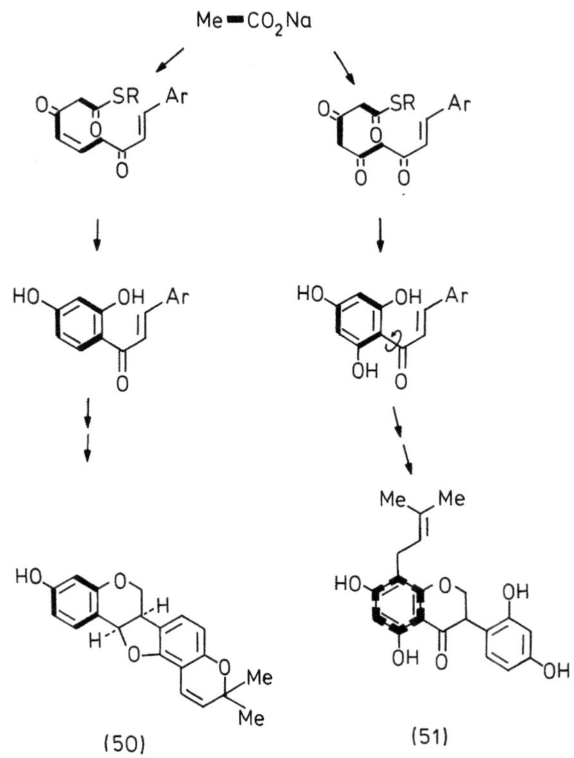

Scheme 8

acetate-enrichment pattern is observed. In contrast, free rotation of a symmetrical chalcone intermediate results in randomisation of label and two labelling patterns in kevitone **(51)**. The biosynthetic routes to these 5-hydroxy- and 5-deoxyisoflavonoids thus diverge prior to the formation of the chalcones as shown in Scheme 8.

Incorporation of [2,3-$^{13}C_2$]phenylalanine into psilotin **(52)** and psilotin epoxide **(53)** by excised shoots of *Psilotum nudum* and analysis of their ^{13}C NMR spec-

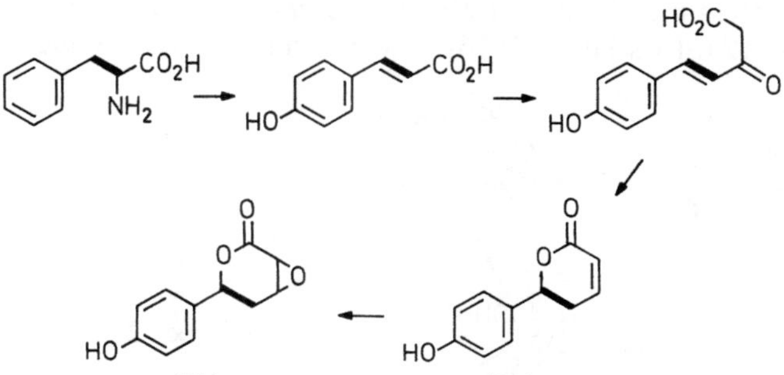

(53) (52)

Scheme 9

tra showed that incorporation of the intact $^{13}C_2$ unit had occurred consistent with their biosynthesis via p-coumaric acid, chain extension with malonate, cyclisation and glucosylation as shown in Scheme 9 (Leete et al. 1982).

Incorporation of [1,2-$^{13}C_2$]acetate by carrot roots treated with ethrel into the phytoalexins 6-methoxymellein **(54)** and eugenin **(55)** confirmed a pentaketide origin for both metabolites. The two modes of acetate incorporation (Scheme 10)

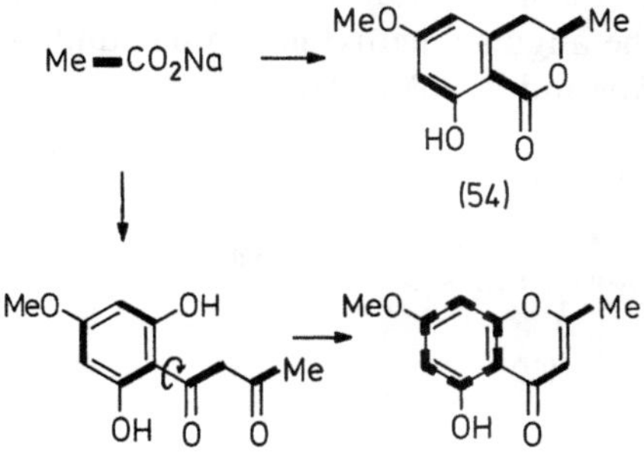

(54)

(55)

Scheme 10

into the benzenoid ring of eugenin indicate that the heterocyclic ring is formed by ring closure onto the two equivalent hydroxyls of a symmetrical intermediate which is not enzyme-bound (Stoessl and Stothers 1973).

A number of studies have been reported with terpenoid metabolites. The labelling pattern of paniculide **(56)** derived from [1,2-$^{13}C_2$]acetate in callus tissue cultures *Andrographis paniculata* demonstrates that the manner of folding of the farnesyl pyrophosphate in the biosynthesis of **(56)** must be as shown in Scheme 11 (Overton and Picken 1976). The sesquiterpenoid stress compounds of the solenaceae have been extensively studied. The labelling patterns observed on incorpo-

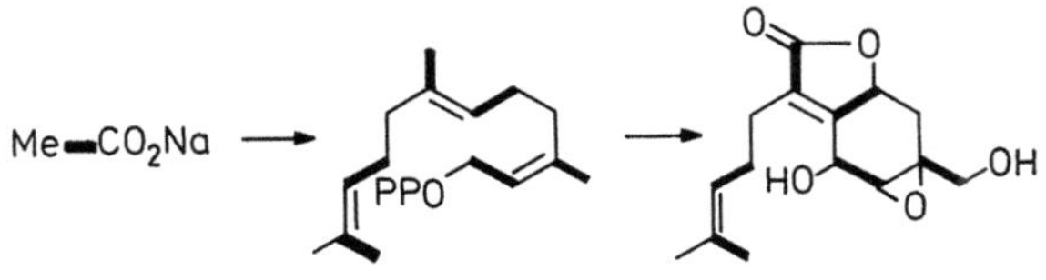

(56)

Scheme 11

ration of [1,2-$^{13}C_2$]acetate into 2,3-germacrenediol **(57)**, lubimin **(58)** and hydroxylubimin **(59)**, the major stress metabolites of *Dataura stramonium* indicate the pathway shown in Scheme 12 (Stoessl et al. 1976). The biosynthesis of the eudes-

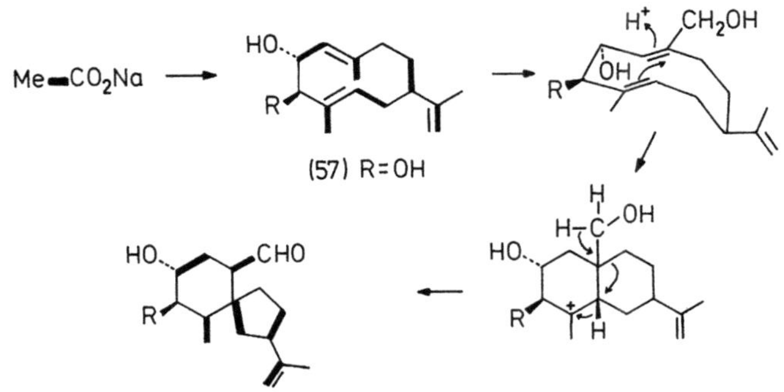

(57) R=OH

(58) R=H
(59) R=OH

Scheme 12

mane sesquiterpenoid, capsidiol **(60)** which is formed by the sweet pepper *Capsicum annum* when it is innoculated with spores of *Monilia fruiticola* has also been studied with [1,2-$^{13}C_2$]acetate. The mode of incorporation of the acetate units suggests a biosynthetic pathway in which the angular methyl group has undergone a 1,2-shift as shown in Scheme 13 (Baker and Brooks 1976).

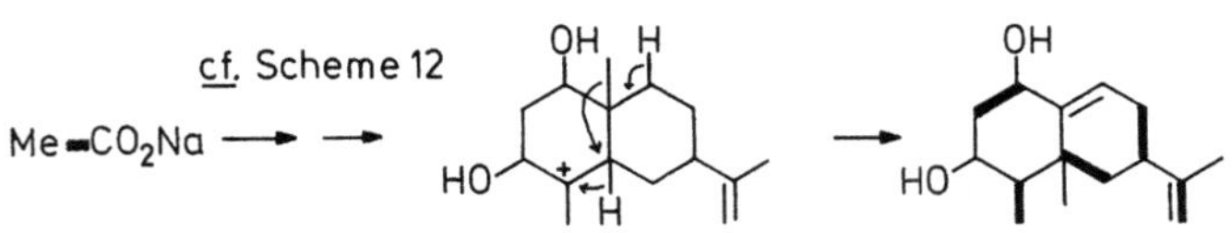

(60)

Scheme 13

Both singly and doubly labelled precursors have been used to study triterpenoid biosynthesis. Incorporation of [4-^{13}C]mevalonic acid into oleanolic acid **(61)**, ursolic acid **(62)** and several other metabolites was achieved using tissue cultures of *Isodon japonicus* (Seo et al. 1975 a). The ^{13}C NMR spectra showed enrichments of carbons 13, 18, and 19 in **(61)** confirming operation of the sequence **(63)**–**(66)** in formation of the oleanes (Scheme 14). The rearrangements involved in ursolic acid biosynthesis could not be identified unambiguously. However, incorporation of [1,2-$^{13}C_2$]acetate gave oleanic acid **(67)** and ursolic acid **(68)** containing a large number of non-adjacent ^{13}C atoms (Seo et al. 1975 b). Although some of these singlets derive from C-2 of mevalonate **(69)**, the absence of coupling between C-20/21 and C-19/29 in **(68)** establishes the pathway to ursane-type triter-

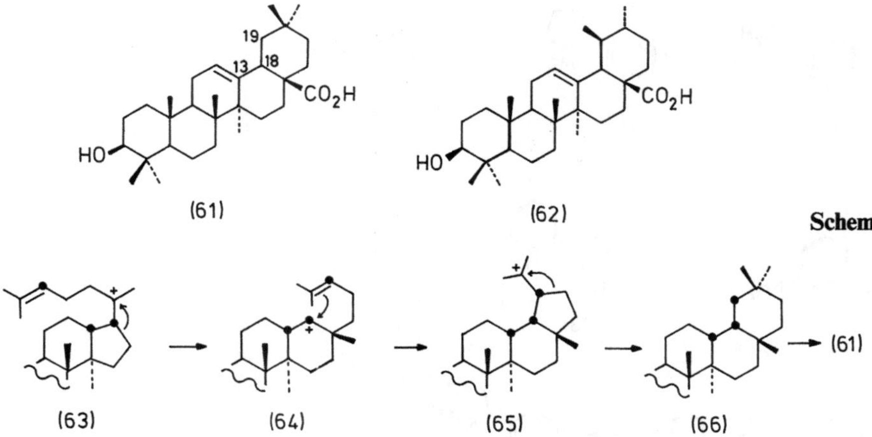

Scheme 14

penoids (Scheme 15) and eliminates alternative skeletal rearrangements such as conversion of **(70)** to **(68)**. Similar studies on a number of triterpenes produced in tissue cultures of a number of plants have been described (Seo et al. 1983).

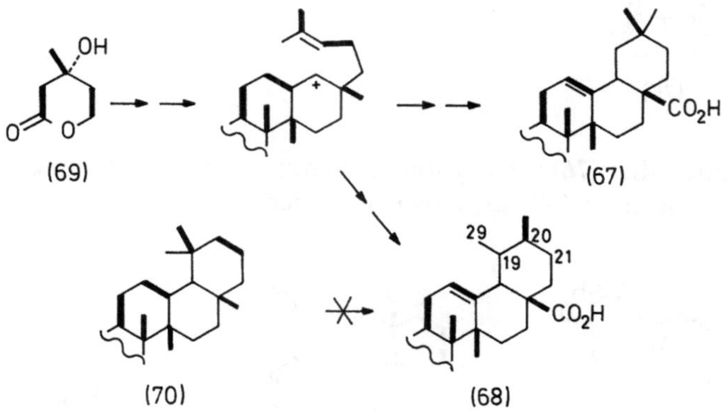

Scheme 15

The technique has also been applied to several alkaloids. [5,6-$^{13}C_2$]Nicotinic acid **(71)** was wick-fed to intact *Nicotinia tabacum* plants. A very low incorporation (0.07%) into nicotine **(72)** was obtained, but even this very low specific incorporation could be detected in the ^{13}C NMR spectrum by the satellites due to the contiguous ^{13}C atoms (Leete 1977).

[4,5-$^{13}C_2$]Lysine was incorporated into anabasine **(73)** by *Nicotania glauca*. The ^{13}C labels were incorporated into C-4′ and C-5′ only (Leete 1982). In contrast, [2,3-$^{13}C_2$]ornithine was incorporated into both C-2′/3′ and C-4′/5′ of nicotine **(72)** to indicate the involvement of a symmetrical intermediate, presumably putrescine **(74)** (Leete and Yu 1980). Incorporation of [2,3-$^{13}C_2$]putrescine into retronecine **(75)** by seedlings of *Senecio isatideus* and observation of coupling satellites on C-1/2 and C-6/7 established the symmetrical labelling of the two halves of the pyrrolizidine alkaloids (Khan and Robins 1981).

Incorporation of [4,5-$^{13}C_2$]lysine into vertine **(76)** by *Heima salicifolia* and analysis of the resultant ^{13}C NMR spectrum showed that C-8 was coupled to

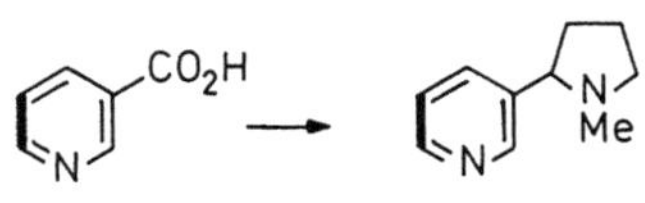

(71) (72)

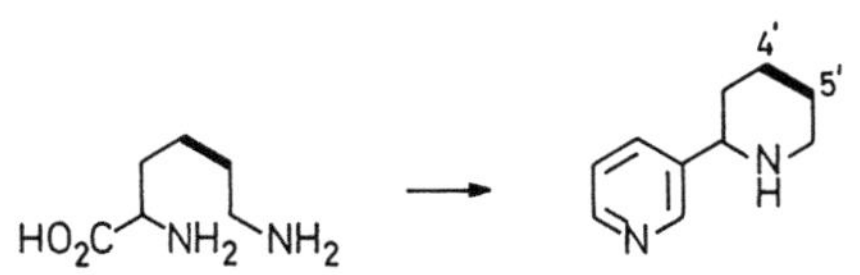

(73)

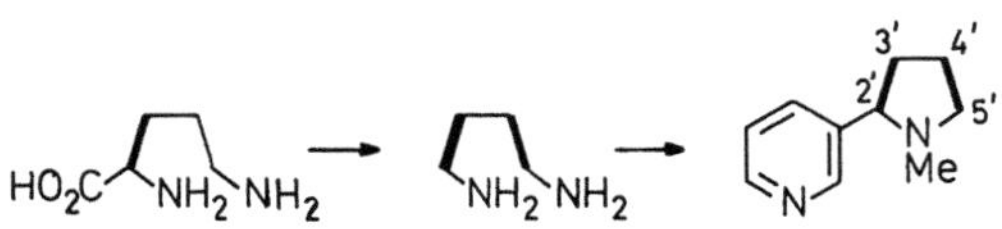

(74) (72)

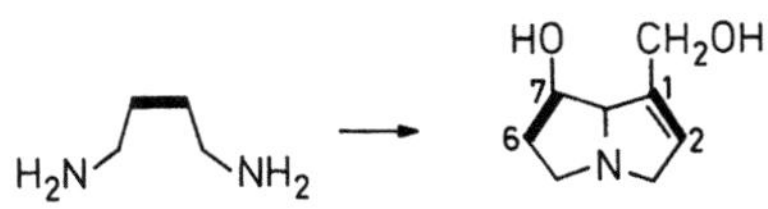

(75)

both C-7 and C-9. It follows that **(76)** is biosynthesised from lysine via a symmetrical intermediate, i.e. cadaverine **(77)** as shown in Scheme 16 (Hedges et al. 1983).

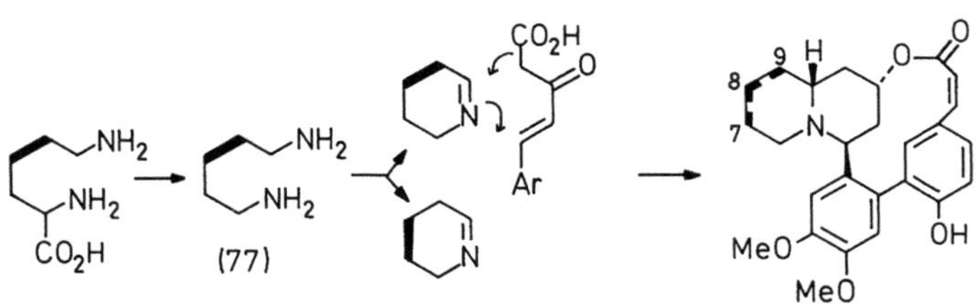

(77) (76)

Scheme 16

Incorporation of [1,2-$^{13}C_2$]acetate into the acridone alkaloid, rutacridone **(79)** using a cell suspension culture of *Ruta graveolens* shows that ring C is derived from three intact acetate units (Scheme 17). The scrambling of labelling however, shows that a symmetrical intermediate e.g. **(78)** must be involved in the pathway (Zschunke 1982).

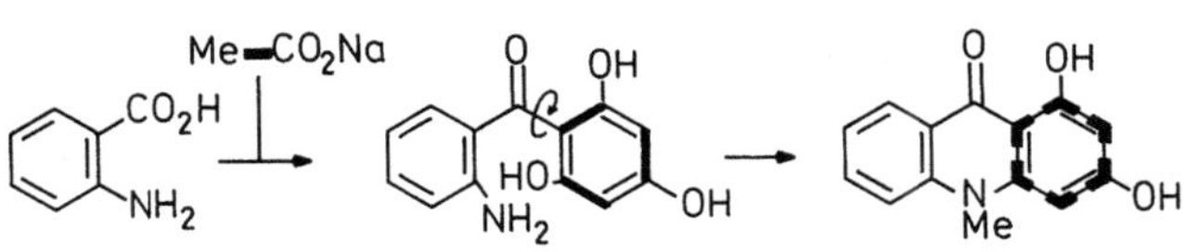

(78) (79)

Scheme 17

6.2.3 Non-Contiguous Double ^{13}C-Labelling

All the above examples in Section 6.2.2 have involved precursors in which there is a one-bond coupling between ^{13}C labels in both precursors and final metabolites. Other studies yielding valuable biosynthetic information have involved precursors with contiguous labels which become non-contiguous in the metabolite, non-contiguous labels which become contiguous, or even non-contiguous labels which remain non-contiguous but detectable during biosynthesis.

6.2.3.1 Two-Bond ^{13}C–^{13}C Couplings from Contiguous $^{13}C_2$-Labelled Precursors

When two ^{13}C-labelled atoms are in a two-bonded relationship to each other, they may or may not display a mutual ^{13}C–^{13}C coupling. These couplings, when observed, are much smaller than one-bond couplings and typically range from 1 to 15 Hz (Marshall 1983). However, with careful spectral determination they can be observed.

In the study of aspyrone **(5)** discussed above, the rearrangement proposed in Scheme 4 involves cleavage of an intact acetate unit during an intramolecular rearrangement. Thus C-1 and C-7 which are in a 2-bonded relationship in aspyrone arise from the same acetate unit. This was confirmed by redetermination of the spectrum of [1,2–$^{13}C_2$]acetate-enriched aspyrone under conditions of higher resolution, whereupon a mutual ^{13}C–^{13}C coupling of 6.2 Hz was observed between C-1 and C-7, Fig. 9c. Similar 2-bond couplings arising from intramolecular rearrangements involving pairs of carbons derived from the same acetate unit have also been observed for vulgamycin **(80)** and the tetrahydrofuran **(81)** (Seto et al. 1975; Seto et al. 1979).

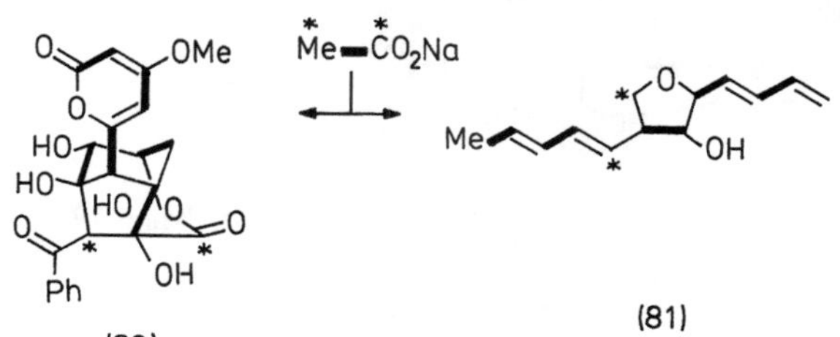

(80) (81)

6.2.3.2 One-Bond ^{13}C–^{13}C Couplings from Non-Contiguous $^{13}C_2$-Labelled Precursors

This is an especially valuable method for elucidating the direction of intramolecular rearrangements and also for establishing the intact incorporation of precursors. It is in essence the reversal of the process described in Section 6.2.3.1 and is best illustrated by examples.

Tropic acid **(82)** as found in scopolamine **(83)** is formed from phenylalanine with the side-chain rearrangement involving an intramolecular 1,2-shift of the

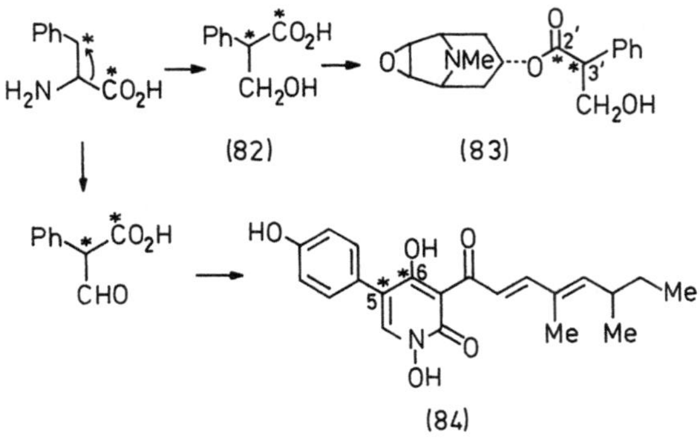

Scheme 18

carboxy-group. This was established by synthesis of [1,3-$^{13}C_2$]phenylalanine and its incorporation into **(82)** and **(83)** in *Datura inoxia* (Leete 1975). In the ^{13}C NMR of the resultant alkaloids, a ^{13}C–^{13}C coupling of 38 Hz was observed between C-2′ and C-3′ to prove that the rearrangement is an intramolecular and not intermolecular process (Scheme 18). Similar studies have been carried out on tenellin **(84)**, an insecticidal metabolite of *Beauveria bassiana*. On incorporation of [1,3-$^{13}C_2$]phenylalanine, the labelled tenellin **(84)** showed a ^{13}C–^{13}C coupling of 45 Hz between C-5 and C-6, indicating that these two carbons have migrated together from C-1 and C-3 of phenylalanine during biosynthesis (Leete et al. 1975).

In a study of cholesterol **(87)** biosynthesis [4,6-$^{13}C_2$]mevalonic acid was fed to a cell-free enzyme preparation from rat liver, C-13 and C-18 showed a ^{13}C–^{13}C coupling of 35 Hz, proving that they were derived from the same mevalonate unit of squalene **(85)** and confirming that C-18 becomes bonded to C-13 by a 1,2-

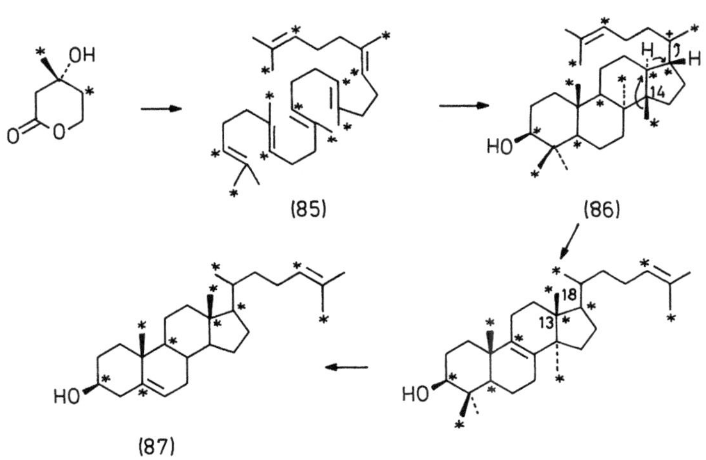

Scheme 19

methyl migration from C-14 of the initial cyclisation product **(86)** rather than by a 1,3 migration from C-8, as indicated in Scheme 19 (Popjak et al. 1977).

One of the first and most elegant examples of this approach was in Battersby's study of uroporphyrinogen III **(89)** biosynthesis. [2,11-$^{13}C_2$]porphobilinogen **(88)** was synthesised and showed a long-range coupling of 4 Hz. This was incor-

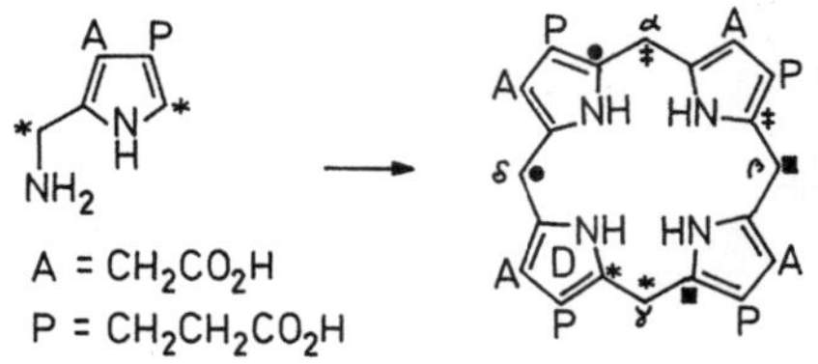

(88) (89)

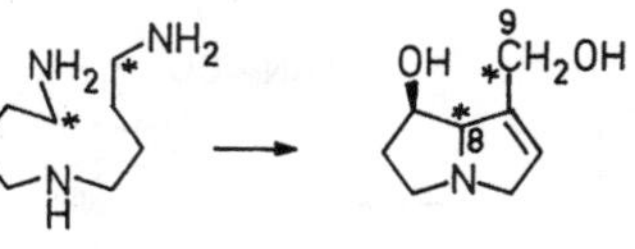

(90) (91)

porated into **(89)** using a cell-free enzyme system from avian blood. Analysis of the resultant ^{13}C NMR spectrum showed three doublets each of 5 Hz splitting corresponding to the α, β and δ carbons, and one doublet of 72 Hz for the γ carbon, indicating that PBG unit D must undergo an intramolecular rearrangement with respect to itself during biosynthesis (Battersby et al. 1973).

6.2.3.3 Two-Bond Couplings from Non-Contiguous $^{13}C_2$-Labelled Precursors

The value of this method lies in its ability to display intact incorporation of complex biosynthetic intermediates. A powerful illustration of the method is the demonstrated intact incorporation of homospermidine into the alkaloid retronecine. [1,9-$^{13}C_2$]Homospermidine **(90)** was synthesised and fed to *Senecio isatideus*. The ^{13}C NMR spectrum of the derived retronecine **(91)** showed doublets for C-8 and C-9 that were superimposed on the natural abundance singlets (Rana and Robins 1983).

The long-range couplings observed for uroporphyrinogen III (see Sect. 6.2.3.2), of course, also demonstrates the intact incorporation of PBG units A, B and C.

7 ^{13}C, ^{15}N Doubly Labelled Precursors

As discussed above, ^{15}N has spin ½ and so will couple to ^{13}C. Thus, observation of a ^{13}C–^{15}N coupling on incorporation of a precursor doubly labelled with ^{13}C–^{15}N can be used to establish intact incorporation of precursor or to probe whether the integrity of a particular carbon-nitrogen bond is maintained during a biosynthetic sequence. ^{13}C–^{15}N Couplings are in general smaller and less predictable than ^{13}C–^{13}C couplings, and in certain cases they may not be observable at all, so results must be interpreted with caution (Levy and Lichter 1979).

As with doubly ^{13}C-labelled precursors we can use precursors in which the ^{13}C and ^{15}N are already in a one-bonded relationship or we can use precursors labelled so that the ^{13}C and ^{15}N become bonded during biosynthesis. An elegant series of experiments encompassing both approaches was carried out during studies on the biosynthesis of the antibiotic streptothricin F **(92)**. As shown in Scheme 20 a series of arginines specifically labelled with ^{13}C and ^{15}N were synthesised and fed to *Streptomyces* L-1689-23. The streptothricin F isolated in each case was analysed by ^{13}C NMR and the ^{13}C–^{15}N couplings indicated by heavy

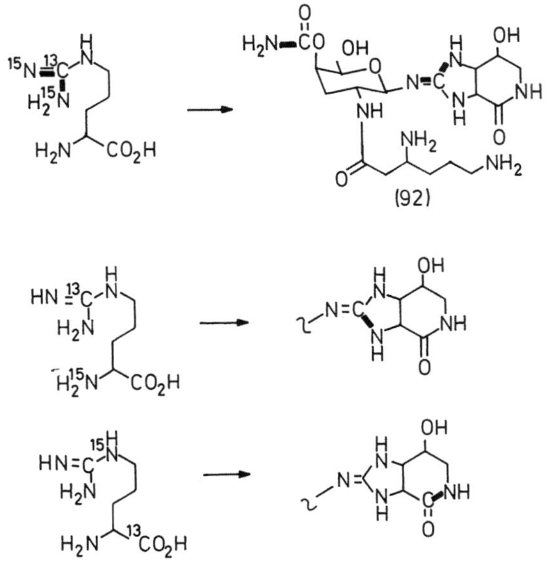

(92)

Scheme 20

lines in Scheme 20 were observed (Martinkus et al. 1983). A particularly elegant application to penicillin biosynthesis has been reported (Baxter et al. 1985). δ-(α-L-Aminoadipoyl)-L-[3-^{13}C]cysteinyl-L-[^{15}N]valine **(93)**, the Arnstein tripeptide, was efficiently incorporated into isopenicillin N **(94)** by a cell-free enzyme system prepared from cells of *Cephalosporium acremonium*. The ^{13}C NMR spectrum showed a ^{13}C–^{15}N coupling of 4.4 Hz for the enriched C-5 resonance of **(94)**.

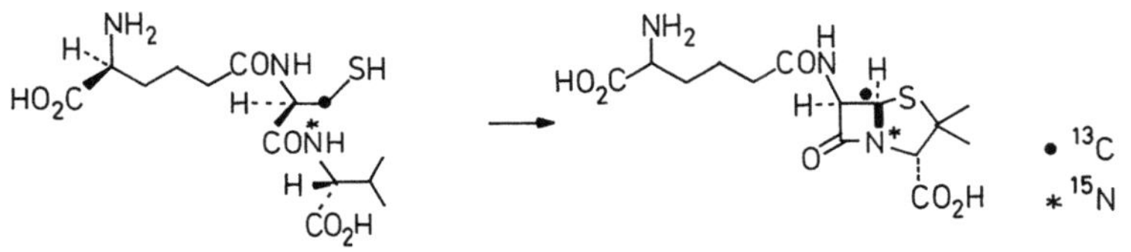

(93) (94)

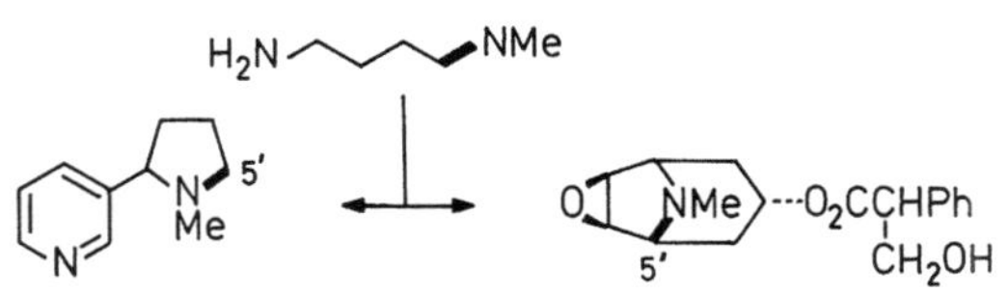

(96) (95)

The technique has also been applied in a number of higher plant alkaloids. [1-^{13}C,methylamino-^{15}N]N-methyl putrescine has been incorporated into scopolamine **(95)** and nicotine **(96)**. In both cases C-5′ show ^{13}C–^{15}N coupling satellites due to the intact incorporation of the contiguous ^{13}C and ^{15}N atoms as shown (Leete and McDonnell 1981). [1-^{13}C,1-^{15}N]Putrescine has been incorporated into retronecine **(98)**. The resultant ^{13}C NMR spectrum had enhanced ^{13}C signals due to C-3, C-5, C-8 and C-9. In addition both C-3 and C-5 showed ^{13}C–^{15}N coupling satellites (4.1 and 3.8 Hz respectively), indicating the intermediacy of a symmet-

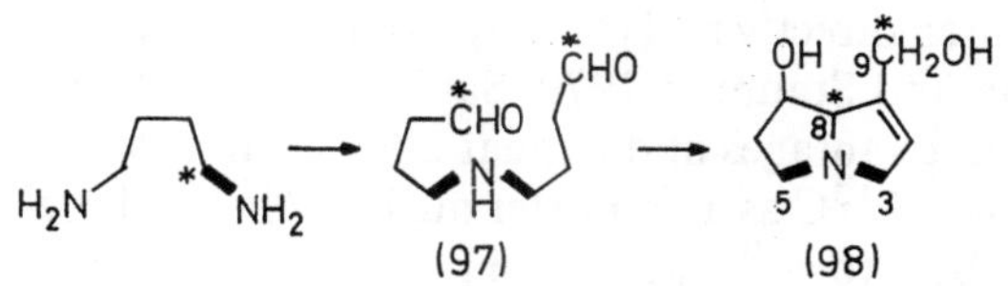

Scheme 21

rical intermediate **(97)** in the biosynthesis of retronecine (Scheme 21) (Khan and Robins 1981).

Similar results have been reported for the quinolizidine alkaloids. [4,5-$^{13}C_2$]Lysine was incorporated into lupinine **(100)** in *Lupinus luteus* plants with the labelling pattern shown in Scheme 22, indicating that a symmetrical intermediate, i.e. cadaverine **(99)** is involved in the biosynthesis of lupinine (Rana and Robins 1984). In further studies [1-^{15}N,1-^{13}C]cadaverine was incorporated into both lupinine **(100)** and sparteine **(102)**. In the spectrum of the enriched lupinine, a ^{13}C–

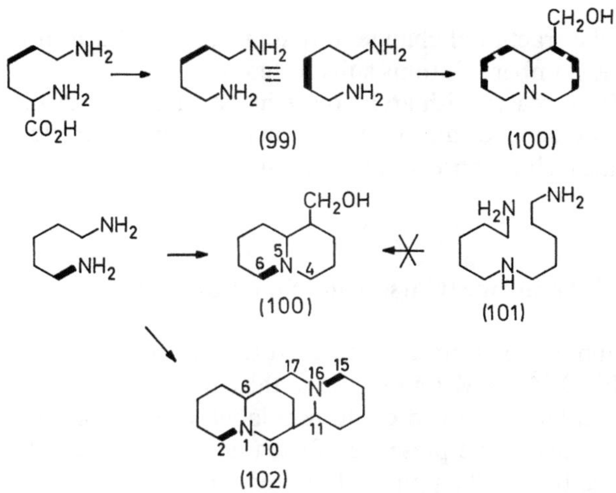

Scheme 22

^{15}N coupling was observed between C-6 and N-5, but not between C-4 and N-5, so that a symmetrical intermediate, e.g. **(101)**, such as is found in the biosynthesis of the pyrrolizidine alkaloids, can be excluded (Golebiewski and Spenser 1983; Rana and Robins 1984). Examination of the ^{13}C NMR spectrum of sparteine **(102)** enriched from [1-^{13}C,1-^{15}N]cadaverine gave results which were similar to those obtained for lupinine. For sparteine, three separate molecules of cadaverine are used and intact incorporation of ^{13}C–^{15}N was observed for C-2 – N-1, and C-15 – N-16 only. Carbons, 6, 10, 11 and 17 appeared as enriched singlets in the ^{13}C NMR spectrum (Rana and Robins 1984; Rana and Robins 1983).

8 Isotope-Induced Shifts in ^{13}C NMR

In studying the nature of the intermediates on a biosynthetic pathway and in particular elucidating the detailed mechanisms of their interconversions, it is essential to determine the biosynthetic origins and fate of the hydrogen and oxygen atoms.

Hydrogen and oxygen can be monitored directly by NMR by using the NMR active isotopes ^{2}H, ^{3}H or ^{17}O (Garson and Staunton 1979; Sankawa et al. 1983). However all of these have distinct disadvantages and a great advance in biosynthetic studies has been made by the use of ^{13}C as a "reporter nucleus" in labelling studies. The presence of ^{2}H alpha or beta to a ^{13}C atom can be deduced from the appearance in the PND ^{13}C NMR spectrum of ^{13}C–^{2}H coupling and/or an isotope-induced shift. Similarly, the presence of ^{18}O alpha to a ^{13}C atom can be detected by an isotope-induced shift in the ^{13}C NMR spectrum. Before discussing specific applications of these methods, some general points on the effect should be noted (Batiz-Hernandez and Benheim 1967).

a) Substitution by a heavy isotope shortens the average length of the bond holding the isotope, and, less so, the remaining bonds of the molecule. This almost always shifts the NMR signal of neighbouring nuclei to high field, although the magnitude of the shift is inversely dependent on the remoteness of the substitution.
b) The shift is largest where the fractional change in mass is largest and is also roughly proportional to the number of atoms substituted.
c) The magnitude of the shift decreases with an increase in "s" character of the bond holding the isotope and depends also on the resonant nucleus, correlating with the range of chemical shifts observed for that nucleus.

8.1 The Deuterium Alpha-Shift Technique (Garson and Staunton 1979)

In this technique, the deuterium label is directly attached to the ^{13}C nucleus in the precursor molecule. The PND ^{13}C NMR spectrum of the enriched metabolite shows, for carbon atoms which have retained deuterium label, a series of resonances upfield of the normal singlet. The presence of each deuterium shifts the centre of the carbon resonance by 0.3–0.6 ppm to low frequency and spin-spin

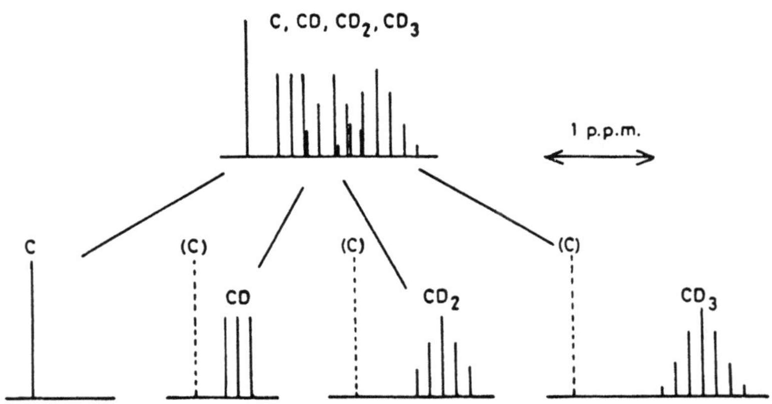

Fig. 11. Expected appearance of the PND ^{13}C NMR spectra of carbons bearing zero, one, two or three deuterium atoms (alpha-shifted)

coupling ($^1J_{CD}$) gives rise to a characteristic multiplet pattern; hence CD appears as a triplet centred 0.3–0.6 ppm upfield of the normal singlet, CD_2 gives a quintet centred 0.6–1.2 ppm upfield and CD_3 gives a septet (Fig. 11). Shifted signals arising from carbons which bear no protium suffer reduced signal-to-noise ratio caused by poor relaxation and lack of NOE enhancement, a disadvantage of the method which is compounded by the multiplicities due to coupling. Deuterium decoupling can assist in this by removing the ^{13}C–^{2}H coupling. However, information not obtainable by direct ^{2}H NMR spectroscopy, such as the distribution of label as CH_2D, CHD_2 and CD_3 and the integrity of carbon-hydrogen bonds during biosynthesis, may be gained.

One of the first applications of this technique was to terrein **(103)**, a cyclopentenone metabolite of *Aspergillus terreus*. In the PND decoupled ^{13}C NMR of terrein **(103)** enriched from [2-^{13}C,^{2}H$_3$]acetate, the signals for isotopically shifted

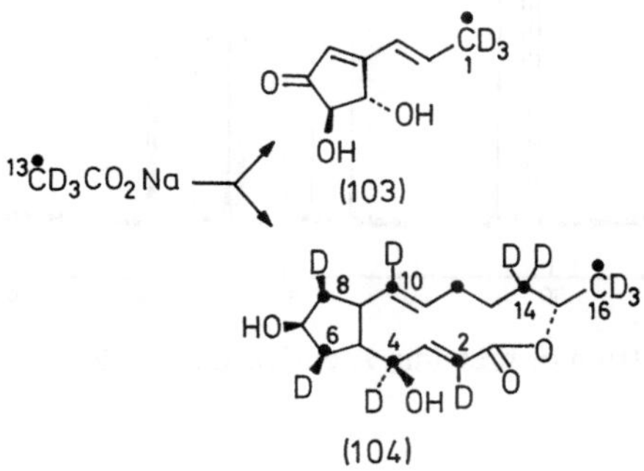

resonances were not clearly observed. However, on redetermining the spectrum with deuterium decoupling, a signal at 17.95 ppm, 0.8 ppm upfield of the normal C-1 signal was observed indicating the presence of molecules triply labelled with deuterium at C-1. In addition a weak doublet signal (J = 123 Hz) centred at 18.2 ppm indicated the presence of molecules labelled as CHD_2. The detection of molecules labelled with three deuteriums at C-1 confirmed that the methyl is derived intact from the methyl of the acetate "starter" unit of the precursor polyketide chain (Garson et al. 1977).

The method has been used most successfully when the ^{13}C NMR spectrum of the enriched metabolite can be determined with simultaneous proton *and* deuterium decoupling (Hutchinson et al. 1981). Thus on incorporation of [2-^{13}C,^{2}H$_3$]acetate into brefeldin A **(104)** by cultures of *Penicillium brefeldianum*, the resultant proton and deuterium decoupled ^{13}C NMR spectrum showed isotopically shifted signals indicating incorporation of up to three deuteriums on C-16, two on C-14 and one on carbons 2, 4, 6, 8, and 10 (Fig. 12).

More recently, a pulse sequence has been described which allows the selective observation of deuterated ^{13}C signals by selective suppression of signals from protonated carbons (Doddrell et al. 1983). This makes the technique more sensitive, but like the simultaneous proton and deuterium decoupling method requires instrumentation and expertise which are not widely available.

The α-shift technique has also found use in terpenoid biosynthesis where it enables hydride shifts to be detected. [3-^{13}C,4-^{2}H$_2$]Mevalonate was incorporated

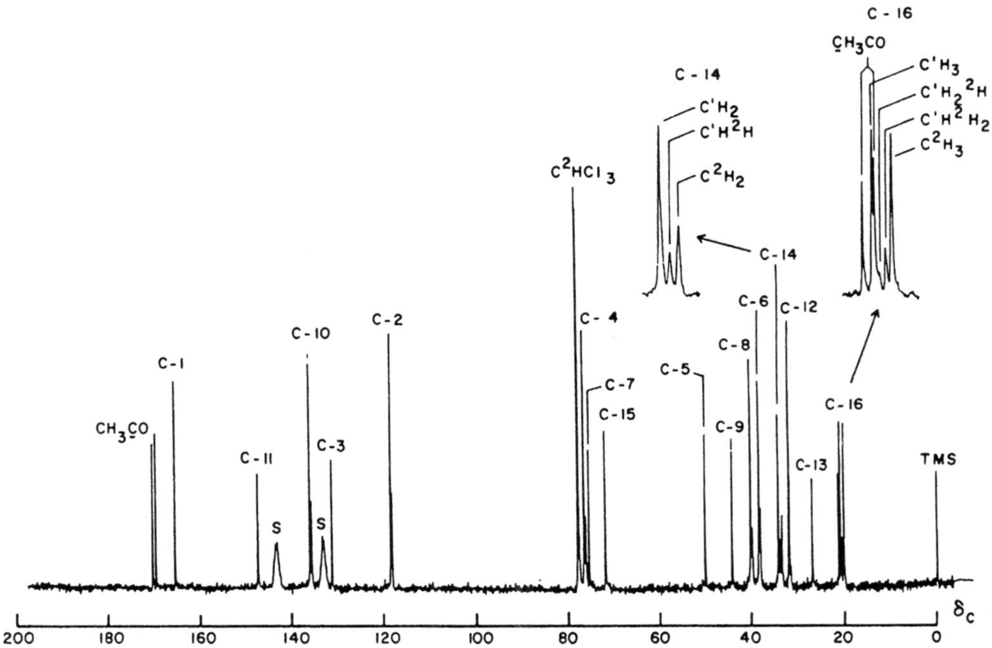

Fig. 12. ^{1}H, ^{2}H-noise decoupled ^{13}C NMR spectrum of brefeldin A enriched from [2-^{13}C, $^2H_3$]acetate

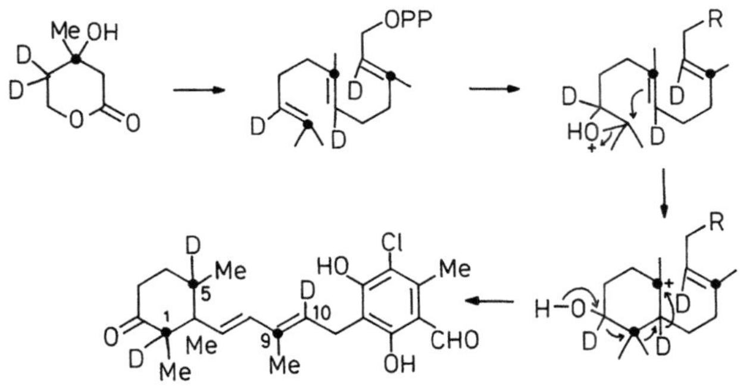

Scheme 23

into the fungal triprenylphenol ascochlorin **(105)**. In the PND ^{13}C NMR spectrum the signal for C-9 was enhanced relative to natural abundance and showed two lines, one for molecules with ^{13}C at C-9 and ^{2}H at C-10 (a β-isotope shift, see Sect. 8.2) and the other, more intense peak for molecules with ^{1}H at C-10. Neither C-1 nor C-5, which are also derived from C-3 of mevalonate, appeared to be significantly enriched relative to natural abundance, consistent with suppression of these signals by attached deuterium. However, two lines of a triplet (J = 24 Hz) centred at 40.65 ppm, were detected upfield of the normal signal for C-5 but no deuterated signal was visible for C-1. Thus the biosynthetic sequence shown in Scheme 22 is indicated (Tanabe and Susuki 1974).

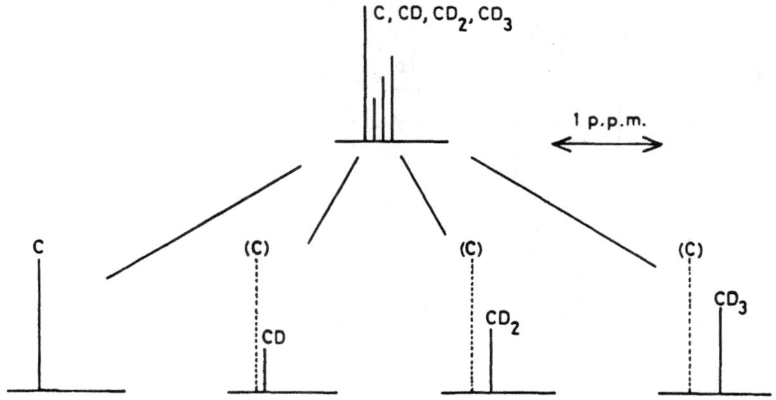

Fig. 13. Expected appearance of the PND ^{13}C NMR spectra of carbons two bonds removed from zero, one, two, or three deuterium atoms (beta-shifted)

8.2 The Deuterium Beta-Shift Technique (Abell and Staunton 1981)

Many of the problems associated with directly attached deuterium are avoided by placing the deuterium label two bonds away from the ^{13}C reporter nucleus. The isotope shift, although reduced, is still observable, and as β-hydrogens only contribute markedly to the relaxation of non-protonated ^{13}C nuclei, the shifted signals otherwise retain any NOE effect also experienced by the unshifted signals on proton decoupling. As geminal carbon-proton coupling constants are generally small anyway (Marshall 1983), and carbon-deuteron couplings are over six times smaller again, the shifted signals are effectively singlets (Fig. 13), even without deuterium decoupling, and this gives a further increase in the signal to noise ratio compared with the corresponding α-shift experiment.

The method was first applied to biosynthesis in a study of 6-methylsalicylic acid **(106)**. [1-^{13}C,$^2H_3$]acetate was fed to cultures of *Penicillium griseofulvum*. The PND ^{13}C NMR spectrum of the methyl ester of the resulting 6-MSA showed shifted signals for C-2, C-4, and C-6, corresponding to the presence of deuterium label at positions 3, 5, and 7 respectively. Thus the integrity of an acetate unit (heavy lines) can be established in certain cases without recourse to a double ^{13}C-labelled experiment (Abell and Staunton 1981). In a similar study on the aflatoxin intermediate averufin **(107)**, the regiospecificity of incorporation of ^{2}H from [1-^{13}C,$^2H_3$]acetate into **(107)** by cultures of *Aspergillus toxicarius* was determined using the β-^{2}H isotope shifts observed in the PND ^{13}C NMR spectrum which is shown in Fig. 14. This shows isotopically shifted signals for C-5′ (the reporter nucleus) indicative of the incorporation of 1–3 deuteriums on C-6′, and for C-1′, C-3′, C-6, and C-8 consistent with the incorporation of one deuterium on C-2′, C-4′, C-5, and C-7 respectively (Simpson et al. 1982).

Unlike the α-shifts, the β-shifts show a marked dependence on both stereochemistry of the deuterium label and the functionality of the ^{13}C reporter nucleus. Carbonyl resonances may show irresolvable or even downfield shifts. However β-shifts do appear to be additive (Simpson and Stenzel 1982).

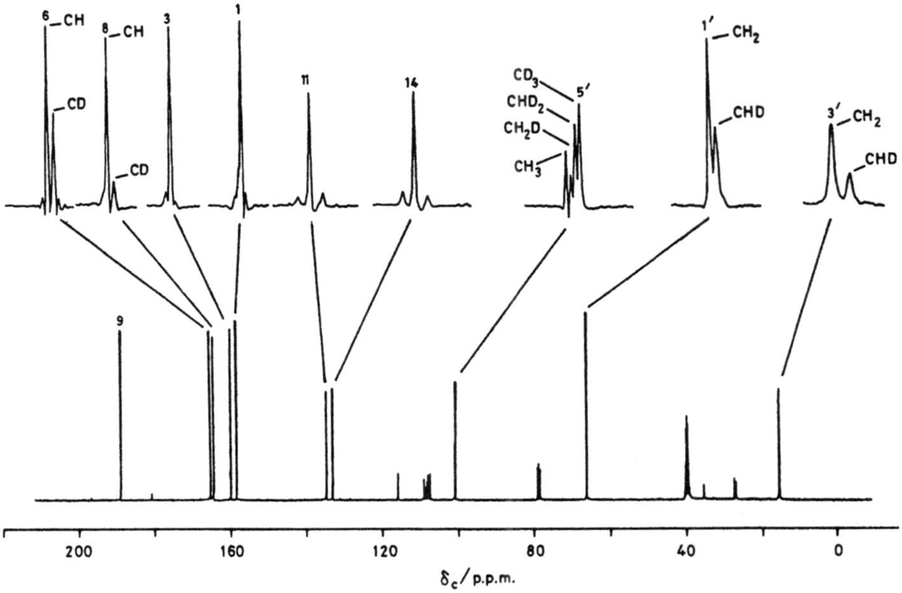

Fig. 14. 90.6 MHz PND ^{13}C NMR spectrum of averufin **(107)** enriched from [1-^{13}C, $^{2}H_3$]acetate

In the only application to date to a higher plant metabolite, the postulated migration of hydrogen from C-5 to C-4 in the biosynthesis of the potato phytoalexins lubimin **(108)**, 3-hydroxylubimin **(109)**, and rishitin **(110)** has been confirmed by the incorporation of [2-^{13}C,$^{2}H_3$]acetate and direct observation of the expected β-shift of the ^{13}C NMR signal of C-5 due to deuterium label at C-4 (Stoessl and Stothers 1983).

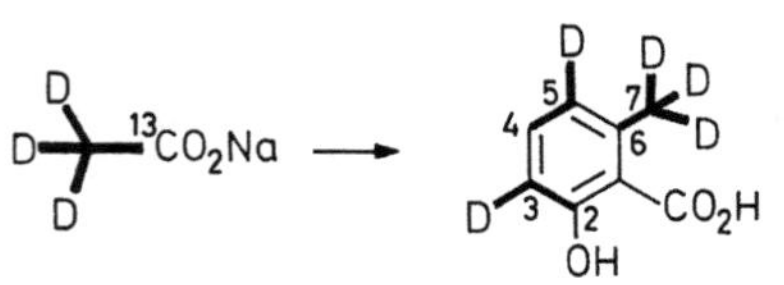

(106)

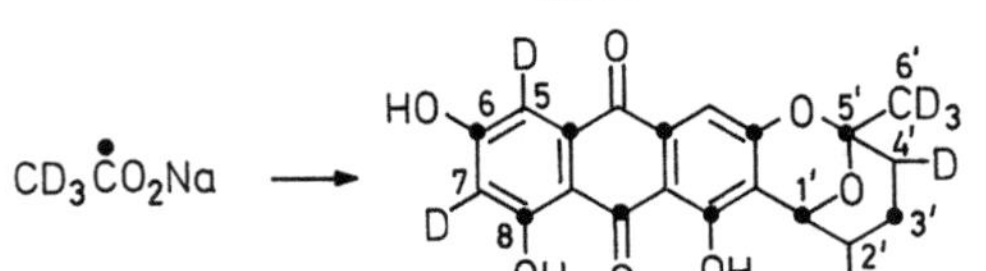

(107)

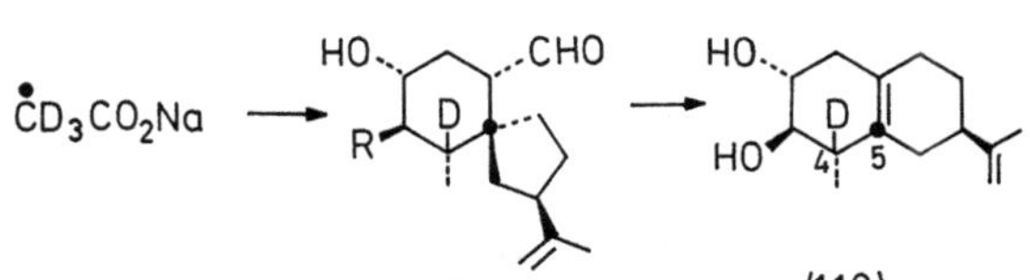

(108) R = H
(109) R = OH

(110)

8.3 ^{18}O Isotope Induced Shifts in ^{13}C NMR

Until recently, the biosynthetic origin or fate of oxygen was determined almost exclusively by mass spectrometry, and locating the position of a label, present perhaps in only low concentrations, was difficult. It was known that the heaviest oxygen isotope, ^{18}O, (natural abundance – 0.204%) was capable of inducing resolvable isotope shifts in the NMR spectra of certain other elements and this had been exploited in biological studies, e.g. the mechanisms and kinetics of enzymatic phosphoryl group transfer had been investigated via the ^{18}O isotope shift in the ^{31}P NMR spectrum (Lowe and Sproat 1978). The detection (Risley and Van Etten 1979) of such an isotope shift in the ^{13}C spectrum provided a more general technique for biosynthetic research, first employed in a study of averufin **(111)**. *Aspergillus parasiticus*, when grown under an atmosphere containing $^{18}O_2$ gas, produced averufin, the ^{13}C NMR spectrum of which showed an isotopically shifted resonance for C-10 only. In an experiment feeding with [1-^{13}C,$^{18}O_2$]acetate, carbons 1, 3, 6, 8, 9, and 1′ all showed prominent shifted signals (Vederas and Nakashima 1980). Thus, at the incorporation levels typically achieved with early precursors, carbon-oxygen bonds which have been preserved intact throughout the course of biosynthesis can be distinguished from those which have arisen between precursor units.

Application of the ^{18}O shift technique has been rapid and the method and its applications have been the subject of a review (Vederas 1982). The shifts show a marked dependence on structure, which could be of use in ^{13}C spectral assignments, but they are not generally much larger than 0.05 ppm. These are very small effects and are the same general magnitude as β-^{2}H isotope shifts. They are only readily observed with high field spectrometers which can determine ^{13}C NMR spectra at ca. 100 MHz.

Although no applications to higher plant metabolites have been reported as yet, the method will certainly find use. One of the more interesting studies reports the incorporation of [1-^{13}C,1,1,4-$^{18}O_3$]-δ-amino-laevulinic acid **(112)** into bac-

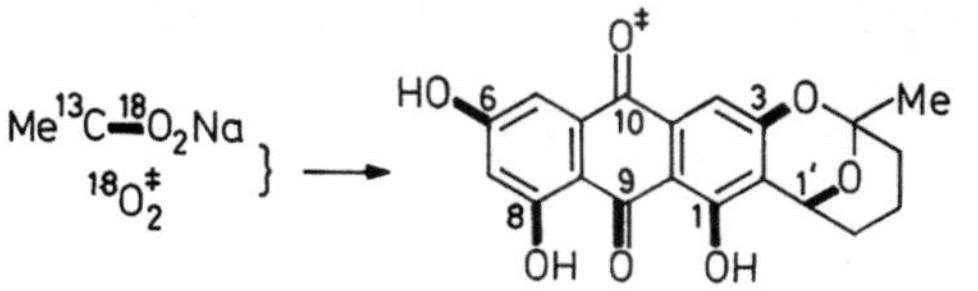

(111)

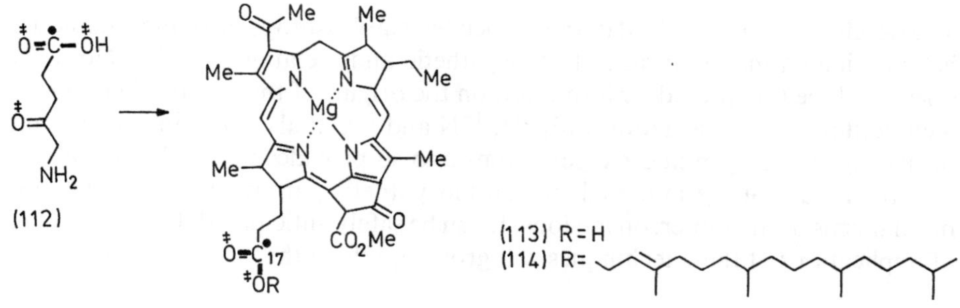

(112)

(113) R=H
(114) R=

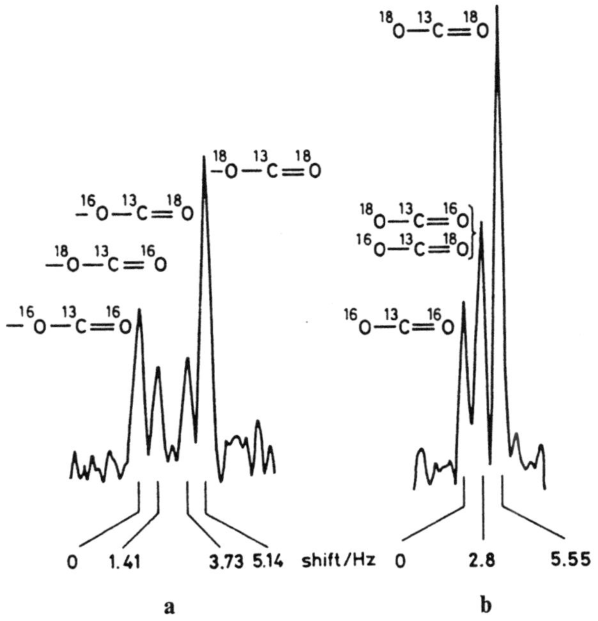

Fig. 15 a, b. Partial PND ^{13}C NMR spectra of: **a** bacteriochlorophyll a **(114)** and **b** δ-aminolaevulinic acid **(112)** illustrating the isotopic distribution at C-17 and C-1 respectively

teriochlorophyll a **(114)** by cultures of *Rhodopseudomonas spheroides*. In the resultant ^{13}C NMR spectrum the signal due to C-17 shows isotopically shifted signals as indicated in Fig. 15, with shifts of 0.014, 0.037 and mainly 0.051 ppm to prove that both the carbonyl and ether oxygens of the phytyl ester linkage are derived from **(112)**. This indicates that phytylation occurs by nucleophilic attack of the C-17 carboxy group of bacteriochlorophyllide a **(113)** on the phytyl pyrophosphate to yield **(114)** with retention of both C-17 oxygen atoms (Emery and Akhtar 1985).

9 Conclusion

The development of ^{13}C NMR and associated stable isotope labelling techniques has provided a major stimulus to biosynthetic and metabolic studies. Not only does ^{13}C labelling provide information on the origins of the carbon skeletons of metabolites but in association with ^{2}H, ^{15}N and ^{18}O it also indicates the origins of hydrogen, nitrogen and oxygen atoms and so provides an invaluable insight into the nature of the intermediates on biosynthetic pathways and the detailed mechanisms of their interconversions. It can be safely anticipated that the number of applications of these techniques will grow rapidly in the next few years.

References

Abell C, Staunton J (1981) The use of ^{2}H Nmr spectroscopy and β-isotopic shifts in the ^{13}C NMR spectrum to measure deuterium retention in the biosynthesis of the polyketide 6-methylsalicylic acid. J Chem Soc Chem Commun 856–858

Abraham RJ, Loftus P (1978) Proton and Carbon-13 NMR spectroscopy. Heyden, London

Ayabe S, Furuya T (1982) Biosynthesis of a retrochalcone echinatin and other flavonoids in the cultured cells of *Glycyrrhiza echinata*. J Chem Soc Perkin Trans 1:2725–2734

Baker FC, Brooks CJW (1976) Biosynthesis of the sesquiterpenoid capsidiol in sweet pepper fruits innoculated with fungal spores. Phytochemistry 15:689–894

Batiz-Hernandez H, Benheim RA (1967) The isotope shift. In: Ensley JW, Feeney J, Sutcliffe LH (eds) Progress in nuclear magnetic resonance spectroscopy. Pergamon Press, Oxford 63–85

Battersby AR, Hunt E, McDonald E (1973) Biosynthesis of type-III porphyrins: nature of the rearrangement process. J Chem Soc Chem Commun 442–443

Battersby AR, Sheldrake PWS, Milner JA (1974) Biosynthesis of colchicine: incorporation of a ^{13}C-labelled precursor in a higher plant. Tetrahedron Letters 3315–3318

Baxter RL, McGregor CJ, Thomson GA, Scott AI (1985) Intact incorporation of δ-(α-L-aminodipoyl)-L-[3-^{13}C]cysteinyl-D-[^{15}N]valine into isopenicillin N. Observation of one-bond ^{13}C–^{15}N coupling. J Chem Soc Perkin Trans 1:369–372

Benn R, Gunther H (1983) Modern pulse methods in high-resolution NMR spectroscopy. Angew Chem Int Ed Engl 22:350–380.

Birch AJ, Simpson TJ (1979) ^{13}C NMR spectra and biosynthesis of phomazarin. J Chem Soc Perkin Trans 1:816–822

Birdsall B, Birdsall NJM, Feeney J (1972) Simplified ^{13}C spectral assignments using a graphical method to present ^{13}C spectra recorded under conditions of proton off-resonance spin decoupling. J Chem Soc Chem Commun 316–317

Bradshaw AP, Hanson JR, Siverns M (1977) Use of induced ^{13}C–^{13}C coupling in terpenoid biosynthesis. J Chem Soc Chem Commun 319

Breitmaier E, Voelter L (1978) ^{13}C NMR spectroscopy. Verlag Chemie, Weinheim

Breitmaier E, Haas G, Voelter W (1979) Atlas of carbon-13 NMR data. Heyden, London

Brown CE, Katz JJ, Shemin D (1972) Biosynthesis of Vitamin B_{12}. Study of carbon-13 nuclear magnetic resonance spectroscopy. Proc Nat Acad Sci USA 69:2585–2588

Brown SA (1972) Methodology. In: Geismann TA (ed) Biosynthesis. The Chemical Society, London (Specialist Periodical Reports, vol 1)

Bu'lock JD (1965) The biosynthesis of natural products. McGraw-Hill, London

Cross BE, Hendley P (1975) The biosynthesis of brefeldin A. J Chem Soc Chem Commun 124–125

Dewick PM, Steele MJ, Dixon RA, Whitehead IM (1982) Biosynthesis of isoflavonoid phytoalexins: incorporation of sodium [1,2-$^{13}C_2$]acetate into phaseollin and kievitone. Z Naturforsch 37C:363–368

Doddrell DM, Staunton J, Laue ED (1983) Use of ^{13}C NMR Spectral editing techniques as a possible procedure to follow the fate of deuterium atoms in mechanistic and biosynthetic studies. J Chem Soc Chem Commun 602–605

Doerfler DL, Ernst LA, Campbell IM (1980) Biosynthetic Intermediates en route to Mycophenolic acid in *Penicillium brevicompactum*. J Chem Soc Chem Commun 329–330

Emery VC, Akhtar M (1985) Mechanistic studies on the phytylation step in bacteriochlorophyll a biosynthesis: an application of ^{18}O induced isotope effect in ^{13}C NMR spectroscopy. J Chem Soc Chem Commun 600–601

Faires RA, Boswell GGJ (1981) Radioisotope laboratory techniques. Butterworths, London

Feline TC, Mellows G, Jones RB, Phillips L (1974) Biosynthesis of hirsutic acid C using ^{13}C nuclear magnetic resonance spectroscopy. J Chem Soc Chem Commun 63–64

Garson MJ, Staunton J (1979) Some new NMR methods for tracing the fate of hydrogen in biosynthesis. Chem Soc Rev 539–561

Garson MJ, Hill RA, Staunton J (1977) Deuterium as a tracer in polyketide biosynthesis: a new method for the detection of chain starter units. J Chem Soc Chem Commun 921–922

Golebiewski WM, Spenser ID (1983) A ^{13}C NMR study of the biosynthesis of lupinine and sparteine. J Chem Soc Chem Commun 1509–1511

Hedges SH, Herbert RB, Wormald PC (1983) Biosynthesis of Lythraceae alkaloids: incorporation of DL-[4,5-$^{13}C_2$,6-^{14}C]lysine and cis and trans-4-(3,4-dihydroxyphenyl)-quinolizidin-2-one into vertine and lythrine. J Chem Soc Chem Commun 145–147

Herbert RB (1981) The biosynthesis of secondary metabolites. Chapman and Hall, London

Holker JSE, Simpson TJ (1981) Carbon-13 nuclear magnetic resonance biosynthetic studies on pentaketide metabolites of *Aspergillus melleus*. J Chem Soc Perkin Trans 1:1397–1400

Holker JSE, Lapper RD, Simpson TJ (1974) Tajixanthone: ^{13}C nuclear magnetic resonance spectrum and feedings whith [1-^{13}C]- and [2-^{13}C]acetate. J Chem Soc Perkin Trans 1:2135–2140

Hutchinson CR, Heckendorf AH, Doddona PE, Hageman E, Wenkert E (1974) Biosynthesis of camptothecin I Definition of the overall pathway assisted by carbon-13 magnetic resonance analysis. J Amer Chem Soc 96:5609–5611

Hutchinson CR, Kurobane I, Mabuni CT, Kumola RW, McInnes IG, Walter JA (1981) Biosynthesis of macrolide antibiotics. 3. Regiochemistry of isotopic hydrogen labelling of brefeldin A by acetate. J Amer Chem Soc 103:2474–2477

Inone K, Ueda S, Nayeshiro H, Inonye H (1982) Biosynthesis of quinones of *Streptocarpus dunii* cell cultures. J Chem Soc Chem Commun 993–994

Inoye K, Shiobara Y, Nayeshiro H, Inonye H, Wilson G, Zenk MH (1979) Site of prenylation in anthraquinone biosynthesis in cell cultures of *Galium mollugo*. J Chem Soc Chem Commun 957–959

Khan HA, Robins DJ (1981 a) Pyrrolizidine alkaloids. Evidence for N-(4-aminobutyl)-1,4-diaminobutane (homospermidine) as an intermediate in retronecine biosynthesis. J Chem Soc Chem Commun 554–556

Khan HA, Robins DJ (1981 b) Pyrrotizidine alkaloid biosynthesis: incorporation of ^{13}C-labelled putrescines into petronecine. J Chem Soc Chem Commun 146–147

Leete E (1977) The incorporation of [5,6-$^{13}C_2$]nicotinic acid into the tobacco alkaloids examined by the use of carbon-13 nuclear magnetic resonance. Bioorg Chem 6:273–286

Leete E (1982) Biosynthesis of anabasine from DL-[4,5-$^{13}C_2$,6-^{14}C]lysine in *Nicotania glauca* examined by ^{13}C NMR. J Nat Prod 45:197–205

Leete E, McDonnell JA (1981) The incorporation of [1-^{13}C,^{14}C,*methylamino*-^{15}N]-N-methylputrescine into nicotine and scopolamine established by means of carbon-13 NMR. J Amer Chem Soc 103:658–662

Leete E, Yu M-L (1980) The incorporation of ornithine-[2,3-$^{13}C_2$] into nicotine acid nor-micotine established by NMR. Phytochemistry 20:1093–1097

Leete E, Kowanko N, Newmark RA (1975 a) Use of carbon-13 nuclear magnetic resonance to establish that the biosynthesis of tropic acid involves an intramolecular rearrangement of phenylalanine. J Amer Chem Soc 97:6826–6830

Leete E, Kowanko N, Newmark RA, Vining LC, McInnes AG, Wright JLC (1975 b) The use of carbon-13 nuclear magnetic resonance to establish that the biosynthesis of tenellin involves an intramolecular rearrangement of phenylalanine. Tetrahedron Letters 4103–4106

Leete E, Ranbom KC, Riddle RM (1979) The metabolism of anatabine to $\alpha\beta$-bipyridyl in *Nicotania* species. Phytochemistry 18:75–78

Leete E, Muir A, Towers CrHN (1982) Biosynthesis of psilotin from [2′,3′-$^{13}C_2$,1′-^{14}C,4-^{3}H]phenylalanine studied with ^{13}C-NMR. Tetrahedron Letters 2635–2638

Levy GC, Lichter RL (1979) Nitrogen-15 nuclear magnetic resonance spectroscopy. Wiley, New York

Levy GC, Lichter RL, Nelson GL (1980) Carbon-13 nuclear magnetic resonance spectroscopy. Wiley, New York

Light RJ, Hahlbrock K (1980) Randomisation of the flavonoid A ring during biosynthesis of kaempferol from [1,2-$^{13}C_2$]acetate in cell suspension cultures of parsley. Z Naturforsch 35C:717–721
Lowe G, Sproat BS (1978) Evidence for a dissociative SNl(P) mechanism of phosphoryl transfer by rabbit muscle pyruvate kinase. J Chem Soc Chem Commun 783–785
Manitto P (1981) Biosynthesis of natural products. Ellis Horwood, Chichester
Marshall JL (1983) Carbon-carbon and carbon-proton NMR couplings. Verlag Chemie International, Deerfield Beach
Martinkus KJ, Tann C-H, Gould SJ (1983) The biosynthesis of the streptolidine moiety in streptothricin F. Tetrahedron 21:3493–3505
McIntyre CR, Simpson TJ (1984) Structural revision and synthesis of LL-D253α: a chromanone metabolite of *Phoma pigmentivora*. J Chem Soc Chem Commun 704–706
Ott DG (1980) Synthesis with stable isotopes. Wiley-Interscience, New York
Overton KH, Picken DJ (1976) Biosynthesis of bisabolene by callus cultures of *Andrographis peniculata*. J Chem Soc Chem Commun 105–106
Popjak G, Edmond J, Anet FAL, Easton NR (1977) Carbon-13 NMR studies on cholesterol biosynthesised from [^{13}C]mevalonates. J Amer Chem Soc 99:931–935
Rana J, Robins DJ (1983a) Pyrrolizidine alkaloid biosynthesis. Intact incorporation of [1,9-$^{13}C_2$]homosperimidine into retronecine. J Chem Res (S) 146–147
Rana J, Robins DJ (1983b) Quinolizidine alkaloid biosynthesis: incorporation of [1-*amino*-^{15}N,1-^{13}C]cadaverine into sparteine. J Chem Soc Chem Commun 1335–1336
Rana J, Robins DJ (1984a) Quinolizidine alkaloid biosynthesis: incorporation of [1-*amino*-^{15}N,1-^{14}C]cadaverine into lupinine. J Chem Soc Chem Commun 81–82
Rana J, Robins DJ (1984b) Biosynthesis of quinolizidine alkaloids. Incorporation of DL-[4,5-$^{13}C_2$,6-^{13}C]lysine into lupinine. J Chem Res (S) 164–165
Risley JH, Van Etten RL (1979) An ^{18}O isotope shift upon ^{13}C NMR spectra and its application to the study of oxygen exchange kinetics. J Amer Chem Soc 101:252–253
Sankawa U, Ebizuka Y, Noguchi H, Isikawa Y, Kitighawa S, Yamamoto Y, Kobayashi T, Itak Y, Seto H (1983) Biosynthesis of citrinin in *Aspergillus terreus*. Incorporation studies with [2-^{13}C,2-$^{2}H_3$], [1-^{13}C,$^{18}O_2$], and [1-^{13}C,^{17}O]acetate. Tetrahedron 21:3583–3591
Seo S, Tomito Y, Tori K (1975a) Biosynthesis of ursene-type triterpenoids from sodium [1,2-$^{13}C_2$]acetate in tissue cultures of *Isodon japonicus* Hara and reassignments of ^{13}C NMR signals in urs-12-enes. J Chem Soc Chem Commun 954–955
Seo S, Tomita Y, Tori K (1975b) Biosynthesis of oleanene- and ursene-type triterpenoids from [4-^{13}C]mevalonic acid in tissue cultures of *Isodon japonicus* Hora. J Chem Soc Chem Commun 270–271
Seo S, Uomori A, Yoshimura Y, Takeda K (1983) Stereospecificity in the biosynthesis of phytosterol side chains: ^{13}C NMR assignments of C-26 and C-27. J Amer Chem Soc 105:6343–6344
Seto H, Sato T, Urano S, Uzawa J, Yonehara J (1976) Utilisation of ^{13}C–^{13}C coupling in structural and biosynthetic studies. The structure and biosynthesis of vulgamycin. Tetrahedron Letters 4367–4370
Seto H, Saito M, Uzawa J, Yonehara H (1979) Utilisation of ^{13}C–^{13}C coupling in structural and biosynthetic studies XII. Biosynthesis of 2-(but-1,3-hydro-4-(penta-1,3-dienyl)-tetrahydrofuran, a metabolite of *Chaetomium coarctatum*. Heterocycles 13:247–253
Simpson TJ (1975) Carbon-13 nuclear magnetic resonance in biosynthetic studies. Chem Soc Review 497–522
Simpson TJ (1979) Carbon-13 nuclear magnetic resonance structural and biosynthetic studies on deoxyherqueinone and herqueichrysin phenalenone metabolites of *Penicillium herquei*. J Chem Soc Perkin Trans 1:1233–1238
Simpson TJ, Stenzel DJ (1982) Application of ^{2}H β-isotopic shifts in ^{13}C NMR spectra to biosynthetic studies. Incorporation of [1-^{13}C,$^{2}H_3$]acetate into O-methylasparvenone in *Aspergillus parvulus*. J Chem Soc Chem Commun 1074–1076
Simpson TJ, de Jesus AE, Steyn PS, Vleggaar R (1982) Biosynthesis of Aflatoxins. Incorporation of [2-$^{2}H_3$]acetate and [1-^{13}C,2-$^{2}H_3$]acetate into averufin. J Chem Soc Chem Commun 632–634

Stoessl A, Stothers JB (1979) The incorporation of [1,2-$^{13}C_2$]acetate into pisatin to establish the biosynthesis of its polyketide moiety. Z Naturforsch 34C:87–89

Stoessl A, Stothers JB (1973) A carbon-13 biosynthetic study of stress metabolites from carrot roots: eugenin and 6-methoxymellein. Canad J Bot 56:2589–2593

Stoessl A, Stothers JB (1983) Biosynthesis of antifungal stress metabolites from potato: observation of hydride shifts via the β-hop from incorporation of [2-$^2H_3$,2-^{13}C]acetate and ^{13}C nmr. Can J Chem 61:1766–1770

Stoessl A, Stothers JB, Ward EWB (1976) Biosynthetic studies of stress metabolites from potatoes: incorporation of sodium acetate-$^{13}C_2$ into sesquiterpenes. Can J Chem 56:645–653. Phytochemistry 15:855–865

Stothers JB (1972) Carbon-13 NMR spectroscopy. Academic Press, New York

Takao N, Iwasa K, Kamijauchi M, Sujiura M (1976) Chem and Pharm Bull (Japan) 24:2859–2865

Takeda R, Katoh K (1983) 3,10-Dihydro-1,4-dimethylazulene a labile biosynthetic intermediate isolated from cultured cells of liverwort *Calyprogeia granulata* Inone. J Amer Chem Soc 105:4056–4058

Tanabe M, Susuki KT (1974) Detection of C-C bond fission during the biosynthesis of the fungal triprenyl-phenol ascochlorin using [1,2-^{13}C]acetate. J Chem Soc Chem Commun 445–446

Tanabe M, Hamasaki T, Suzuki Y, Johnson LF (1973) Biosynthetic studies with carbon-13: Fourier transform nuclear magnetic resonance spectra of the metabolite avenaciolide. J Chem Soc Chem Commun 212–213

Vederas JC (1982) Biosynthetic studies using ^{18}O isotope shifts in ^{13}C nuclear magnetic resonance. Can J Chem 60:1637–1642

Vederas JC, Nakashima TT (1980) Biosynthesis of averufin by *Aspergillus parasiticus*: detection of ^{18}O-label by ^{13}C NMR isotope shifts. J Chem Soc Chem Commun 183–185

Wasserman HH, Sykes RJ, Shaw CK, Cushley RJ (1974) The biosynthesis of metacycloprodigiosin and undecylprodigiosin. Tetrahedron Letters 2787–2791

Zschunke A, Baumert A, Gröger D (1982) Biosynthesis of rutacridone in cell cultures of *Ruta graveolens*. Incorporation studies with [^{13}C]acetate. J Chem Soc Chem Commun 1263–1265

Determination of the Energy Status of Plant Cells by ^{31}P-Nuclear Magnetic Resonance Spectroscopy

J. K. M. Roberts

1 Introduction

Metabolism involves the coupling of the energy of light, or the energy released from catabolism, to synthesis, transport, and movement. This coupling is achieved primarily via ATP (Lipmann 1941). Metabolism of ATP is thus synonymous with energy metabolism.

There is interest currently in the factors determining the rate at which ATP is synthesized in vivo, relative to rates of reactions that consume ATP (Atkinson 1977; Erecinska and Wilson 1982; Pradet and Raymond 1983). A close link between these two types of reaction is apparent, because the concentration of ATP in cells changes relatively little over a wide range of metabolic conditions (Pradet and Raymond 1983). A close link might also be deemed essential, since the ATP in aerobic plant tissue turns over every few seconds (Roberts et al. 1984 b).

Three principal aspects of energy metabolism require examination at present. First, the concentration of free ATP in the various intracellular compartments must be determined in cells under different physiological conditions. This is useful in respect to the affinities of ATP-dependent enzymes, for the concentration of MgATP might be important in determining enzyme activity, and so perhaps metabolic rates. Second, the ability of this MgATP to perform useful work (work that could drive metabolism) must be determined under different physiological conditions. This property of ATP in cells is determined not only by the concentration of MgATP, but also by the concentrations of products of MgATP breakdown: ADP, AMP, P_i, PP_i, and H^+. Third, rates of the various reactions responsible for ATP synthesis and breakdown must also be determined. Once this information is available, the relative importance of thermodynamic (Wilson et al. 1974; Erecinska and Wilson 1982), allosteric (Atkinson 1977; Pradet and Raymond 1983) or other factors determining rates of ATP synthesis and/or utilization might be appreciated.

In this chapter, the use of ^{31}P-NMR to measure concentrations of free ATP and ADP, and the Gibbs' free energy for hydrolysis of ATP to ADP, P_i and H^+ in living plant tissues is described. The use of ^{31}P-NMR to measure rates of ATP synthesis in the steady state, in vivo, is described in detail elsewhere (Roberts et al. 1984 b).

2 Principles

When a plant tissue is placed in a modern, high field Fourier transform NMR spectrometer that is set up to observe ^{31}P, a spectrum containing a number absorption lines at different frequencies may be obtained (e.g., Fig. 1). Each absorption line arises from subpopulations of ^{31}P nuclei in the sample that are (1) in distinct chemical environments, determined by the nature of the chemical group and other chemical conditions (ionic strength, pH, divalent ion concentration); and also (2) present at a tissue concentration greater than ca. 0.1 mM; and (3) are freely mobile (correlation time greater than that of large proteins). So, in ^{31}P-NMR spectra of aerobic plant tissues, separate signals are detected from each of the three phosphate moieties of ATP (α, β, and γ); and these signals occur in a different region of the spectrum compared to signals from intracellular inorganic phosphate, which are in turn separated from signals from monophosphate esters, such as glucose-6-phosphate (Fig. 1).

In addition to identifying the existence of such compounds in living tissue, ^{31}P-NMR can be used to determine the concentrations of such compounds, for the intensity of a given NMR peak is linearly proportional to the concentration of nuclei responsible for the particular peak. Finally, the positions (or "chemical shifts") of some spectral peaks can provide information on certain intracellular conditions. For example, separate signals from inorganic phosphate in cytoplasm and vacuoles are observed (Fig. 1), because the chemical shift of P_i is very pH-dependent, and because there is a large ΔpH between cytoplasm and vacuole (Roberts et al. 1980). Another important example is the change in the chemical shift of the ATP ^{31}P resonances on binding magnesium ion; this is discussed below.

These are the attributes of in vivo ^{31}P-NMR that may be useful for the study of energy metabolism. The theory of NMR spectroscopy is covered in detail in

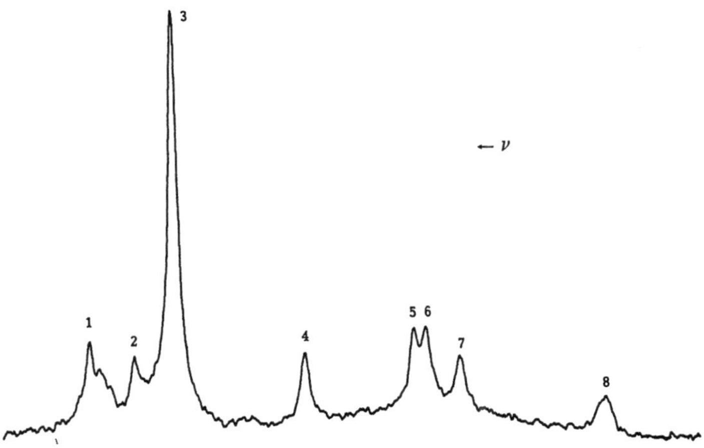

Fig. 1. ^{31}P-NMR spectrum of etiolated pea stems. The tissue was perfused with oxygenated 15 mM sucrose at 40 ml min^{-1}. *Arrow* indicates the direction of increasing frequency. Peak assignments: *1* glucose-6-phosphate; *2* cytoplasmic inorganic phosphate; *3* vacuolar inorganic phosphate; *4* γNTP; *5* αNTP; *6, 7* uridine diphosphoglucose; *8* βNTP

various monographs (Farrar and Becker 1971; Shaw 1976; Becker 1980), and the general topic of in vivo NMR spectroscopy is the subject of numerous reviews (e.g., Gadian and Radda 1981; Gadian 1982; Shulman et al. 1979; Scott and Baxter 1981; Roberts and Jardetzky 1981; Roberts 1984).

3 Experimental Considerations

NMR spectroscopy is an intrinsically insensitive technique. At present, accurate quantitation of levels of phosphates is possible only when their concentration exceeds ca. 0.1 mM, and after attention is paid to the factors that determine experimental sensitivity. Here the factors contributing to sensitivity in NMR spectroscopy are briefly described. This is followed by a description of the procedures used in this laboratory to maintain plant tissues under constant and defined physiological conditions in the NMR spectrometer.

Sensitivity in NMR experiments is determined by: (1) the field strength of the spectrometer magnet. As described by the Boltzmann distribution (see Becker 1980), in a given sample, more nuclei are observed by NMR spectroscopy at higher magnetic field strengths. More powerful NMR magnets are less common, and more expensive. For most quantitative NMR studies of plants, spectrometers operating at frequencies greater than ca. 120 MHz for ^{31}P (300 MHz ^{1}H) are desirable, if not essential. In the context of the subject of this chapter, such high field instruments are also necessary in order to resolve ^{31}P-NMR signals from γATP and βADP (discussed below). (2) The volume of tissue over which NMR signals are detected. While it is not possible or desirable to increase tissue concentrations of metabolites detectable by NMR, the total number of nuclei in the sample can obviously be increased by increasing the sample volume. This is an important point, since most spectrometers (used by chemists) have very small sample volumes (5 or 10 mm diameter, and ca. 20 mm long). Increasing the sample volume requires building new sample probes, using components costing a few dollars. Sample diameters of 20 mm are practicable in many spectrometers. (3) The density of cells in the sample volume. It is desirable to pack the biological material into the sample chamber as densely as possible, in order to maximize signals. Limits to the density are determined by the conflicting requirements for perfusion of tissues (see below), or bubbling of cell suspensions (Martin et al. 1982). (4) The values of particular parameters in the NMR experiment, principally the length of the excitation pulse, and the delay between the exciting pulses (see below). The effects of these parameters on the ratio of spectral signal/noise are discussed in detail elsewhere (Shaw 1976). (5) Numerous parameters involved in spectrometer design, construction, and operation. Most obvious among these are sample probe design and tuning, magnetic field homogeneity over the sample volume, and protection of detected signal from noise by filters, gates, and shielding. These are mentioned here principally because problems in these areas can be both significant causes of low sensitivity, and difficult or laborious to rectify. The problem is worse when spectrometers are used for many purposes (particularly if switches between the different resonance frequencies of different nuclei are commonly made). The absence of accepted sensitivity standards for in vivo NMR, as exists for chemical applications, together with the great heterogeneity among different plant tissues, makes it difficult to know when spectrometer sensitivity is exceptional, or substandard. This is particularly true early in an in vivo project, when the ability to get an NMR signal at all may be in doubt.

Because high densities of cells in the NMR sample volume are required to maximize signal intensity, it is essential to ensure that the metabolic needs of the sample are met. The sample must be maintained under defined and homogeneous physiological conditions for in vivo NMR experiments to yield useful results. In this laboratory we employ a perfusion system, suitable for study of almost any plant tissue by in vivo NMR. The elements of this system are illustrated in Fig. 2. The key features are, first, a specially built probe permitting access for tubing from both above and below the sample; second, a reservoir-sample-pump arrangement, not involving special precision connections, that permits very high flow rates without the danger of pressure buildup if the circuit clogs for some reason. The temperature of the sample is determined by the temperature of the perfusion medium. The probe has no provision for field frequency locking (see Becker 1980), since this is unnecessary for in vivo experiments using a super-conducting magnet, even for experiments lasting several days. Neither is proton decoupling possible with this probe, since this too is of no value for in vivo ^{31}P-NMR studies. The homogeneity of the magnetic field is optimized ("shimmed") over the sample volume by maximizing the ^{31}P-NMR signal from concentrated

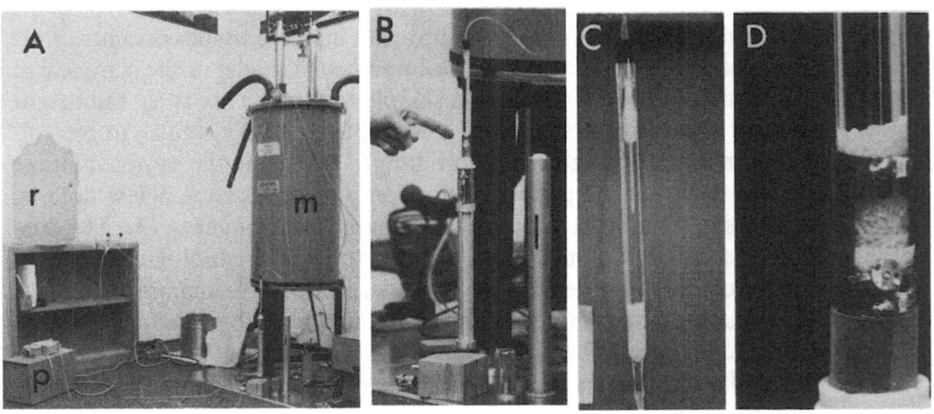

Fig. 2 A–D. Apparatus for study of perfused plant tissues by NMR. **A** perfusion route begins with a 50-l reservoir containing perfusion medium (*r*), tubing passes into the top of the magnet (*m*) to the sample (here shown outside the magnet), and then to a high volume peristaltic pump (*p*) (Harvard Apparatus, South Natick, MA), which pulls the medium through the sample. **B** modified NMR probe, with rf shield removed (cylinder on *right*) that permits access to the detecting coil (indicated by *finger*) from above (*inlet*) and below (*outlet*). **C** close-up of sample tube, 12 mm in diamter showing a coarse glass wool plug supporting the sample. **D** close-up of a root tip sample in the ^{31}P-NMR coil, built by Dr. David Wemmer at Stanford; the central capillary containing methylene diphosphonate as a reference compound is visible

potassium phosphate, contained in a tube of the same dimensions as that used to contain the tissue. Once optimized, we find no need to change the shims if different samples are used, even over several days of experiments. We simply ensure that the pieces of tissue are packed for the most efficient use of the limited sample volume (ca. 2.5 cm^3), and that air bubbles in the sample volume are eliminated (see Chap. 5, this Vol. for a discussion of the effect of air bubbles on spectral quality). This shimming routine is much simpler than the general approach used in most NMR studies of animal tissues, which we previously employed (Roberts et al. 1982), whereby the signal from protons of tissue water is maximized to optimize magnetic field homogeneity (Gadian 1982).

There are a number of advantages associated with such a perfusion system. The environment about the tissue can be kept absolutely constant indefinitely; no buildup of products of metabolism (e.g., H^+, ethanol), or depletion of nutrients (sugars, oxygen) occurs. Or, if desired, conditions can be changed extremely rapidly, and uniformly. For example, the transition from normoxia to anoxia can be completed within 2 min (Roberts et al. 1984a), the bulk of this transition period being due to the time taken for tissue oxygen to be fully depleted such that oxidative phosphorylation ceases. A further advantage of perfusion is that metabolic phenomena, other than those observable using NMR, can be followed simultaneously with NMR spectroscopy, simply by monitoring the effluent from the sample. These include oxygen and carbon dioxide exchange between the sample and perfusate, H^+ extrusion, and uptake or release of ions or organic molecules.

Using this experimental arrangement, on our modified Bruker HXS-360 NMR spectrometer, we can observe ATP in normoxic maize root tips in 10 min. Cytoplasmic pH and the concentration of cytoplasmic inorganic phosphate can be quantitated ca. every 1 min. The concentration of ADP in plants in usually much lower than that of ATP or inorganic phosphate, requiring signal averaging over periods of one to several hours for reliable quantitation.

4 Quantitation of Metabolite Concentrations in vivo by NMR

As noted above, the intensity (i.e. area) of a particular NMR spectral peak is linearly proportional to the amount of the chemical group responsible for the peak, in the sample volume. Hence, concentrations of compounds observable by NMR can be determined by integration of peak areas. In order to do this, it is first necessary to determine the magnitude of the proportionality constant between peak area and concentration. Two factors determine the magnitude of this constant. First, the sensitivity of the spectrometer will influence peak intensity. Sensitivity is calibrated using a reference compound included in the sample volume, in a manner analogous to use of internal standards in chromatographic methods. The area of a peak in a spectrum of plants can be compared to the peak area from a solution of a compound at a known concentration. The comparison is made by including the same reference capillary in both the plant sample and the standard solution. In addition, to obtain tissue concentrations by NMR, it may also be necessary to correct for the proportion of the sample volume not occupied by tissue.

Second, the proportionality constant that links peak area and concentration often differs from peak to peak. Accurate quantitation is not possible before this effect is accounted for. This complication can arise because of the way in which NMR spectra are acquired. NMR spectra are acquired by adding together many signals from nuclei obtained over periods ranging from ca. 1 min to several hours. The signals are accumulated in a cyclic sequence consisting of: excitation pulse,

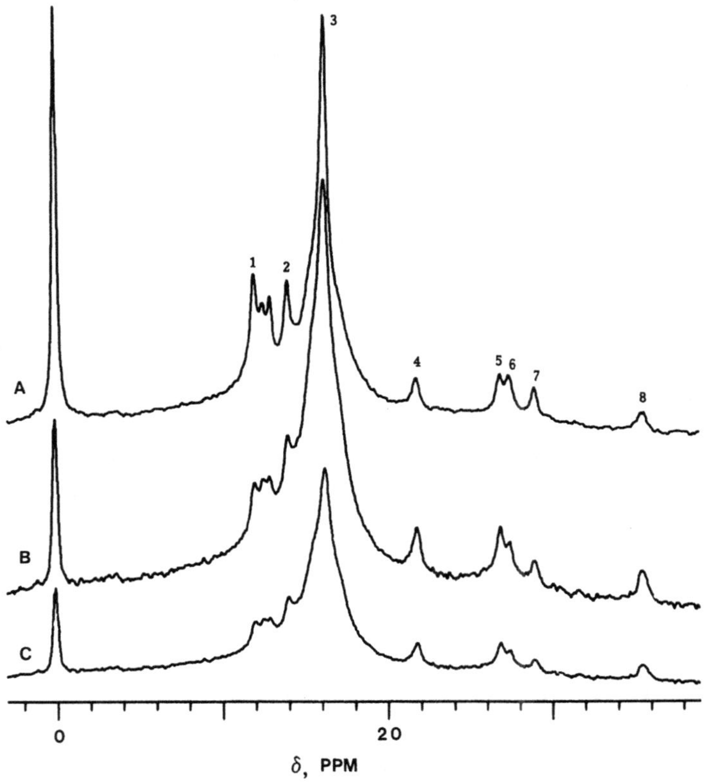

Fig. 3. Effect of pulse repitition rate on the relative intensities of resonances, and on spectral signal-to-noise. Spectrum *A* was obtained in 200 min, with a pulse interval of 10 s (^{31}P spins essentially completely relaxed before each exciting pulse). Spectrum *B* was obtained in 20 min, with a pulse interval of 0.5 s (^{31}P-spins partially saturated before each exciting pulse). The spectra are of the same sample, and were obtained over the same time period (alternating accumulation). The relative intensities of peaks *1* (glucose-6-phosphate) and *6* and *7* (uridine diphosphoglucose) are reduced in intensity relative to the other peaks in spectrum *B*, compared to *A*. Spectrum *C* is a re-plot of spectrum *B*, such that the spectral noise level is the same as in spectrum *A*. It is evident that the signal from nucleotides (peaks *4*, *5*, and *8*) is nearly as strong in spectrum *C* as in spectrum *A*, even though spectrum *C* was obtained in one tenth the time used to acquire spectrum *A*. The peak at 0 ppm is due to methylene diphosphonate, used as an external reference. For other peak assignments see legend to Fig. 1

collect signal, recovery period, repeat. The recovery period is included in this sequence to allow for a sufficient number of nuclear spins to return to the ground energy state for a signal to be observed following the next exciting pulse. If the exciting pulses are separated by too short an interval, little or no signal is observed, and the population of nuclear spins in the sample is said to be "saturated". The ratio of signal to noise in NMR spectra is maximal (with respect to time taken to acquire the spectra) when they are acquired under partially saturating conditions. Such conditions are commonly used for in vivo NMR studies, since every possible means of improving spectral quality is often required to obtain useful spectra. However, the rates at which nuclear spins in different chemical groups recover after excitation (specifically the rate of "longitudinal relaxation") may differ. Hence, under conditions of partial saturation, signals from chemicals that relax quickly after excitation will be disproportionately stronger than signals which recover more slowly. The consequences of saturation are illustrated in Fig. 3. Certain signals, notably a broad component of the vacuole inorganic phosphate resonance, and the nucleoside triphosphate resonances, are favored relative to others by the rapid pulse experiment. Peaks from sugar nucleotides and, particularly, glucose-6-phosphate, on the other hand, show strong effects of saturation. Figure 3 also illustrates the benefits of acquiring spectra under conditions of partial saturation: signals from nucleoside triphosphates of intensity comparable to those observed under nonsaturating conditions can be obtained ten times more quickly. Intensities of peaks in spectra obtained under saturating conditions can be related to concentrations by applying correction factors. The correction factors are determined by comparing relative peak areas in spectra obtained under saturating and nonsaturating conditions (Dawson et al. 1977).

Often, changes in the concentrations of phosphates observable by ^{31}P-NMR occur slowly, and so the need to acquire spectra as quickly as possible is diminished. In such circumstances (e.g., Roberts et al. 1984c), it may be possible to acquire all spectra under nonsaturating conditions. This greatly simplifies data analysis.

5 Measurement of Nucleoside Di- and Triphosphate Levels by ^{31}P-NMR

A spectrum of a mixture of AMP, ADP, and ATP (in the presence of excess magnesium ion) is shown in Fig. 4. All the different nucleoside triphosphates give the same ^{31}P-NMR spectrum, consisting of peaks attributable to each of the phosphate groups: α, β, and γ. ^{31}P-NMR spectra of nucleoside diphosphates give two lines, with chemical shifts very similar to those from α and γNTP. AMP gives only one ^{31}P-NMR signal, resonating approximately 0.7 ppm upfield from the glucose-6-phosphate signal, at pH 7.6.

Because line-widths of in vivo ^{31}P resonances are relatively broad, it is only at relatively high magnetic field strengths (greater than ca. 6.9 Tesla: 129 MHz ^{31}P) that resolution of the γNTP can be resolved from the βNDP line (Fig. 4). The

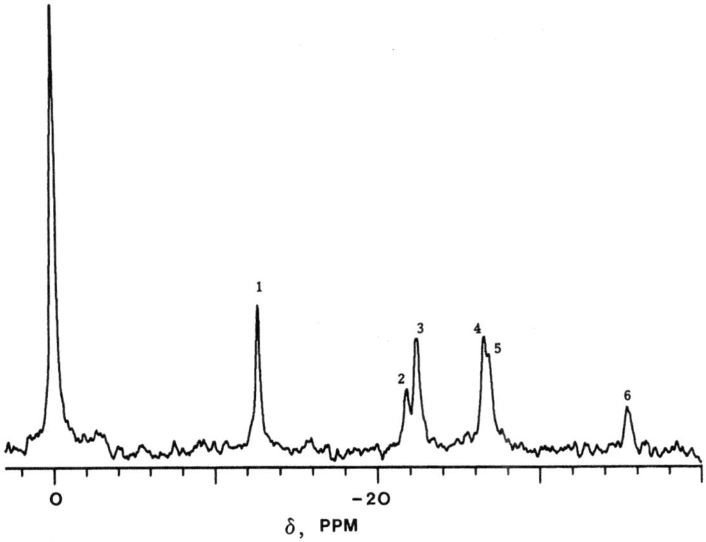

Fig. 4. 145.7 MHz ^{31}P-NMR spectrum of a mixture of AMP, ADP and ATP (0.5–1 mM). The solution also contained 50 mM KCl, 2 mM $MgCl_2$ and 10 mM Tris base, brought to pH 7.6 with MES. Peak assignments: *1* AMP; *2* γATP; *3* βADP; *4* αADP; *5* αATP; *6* βATP. The peak at 0 ppm is due to methylene diphosphonate, used as an external reference

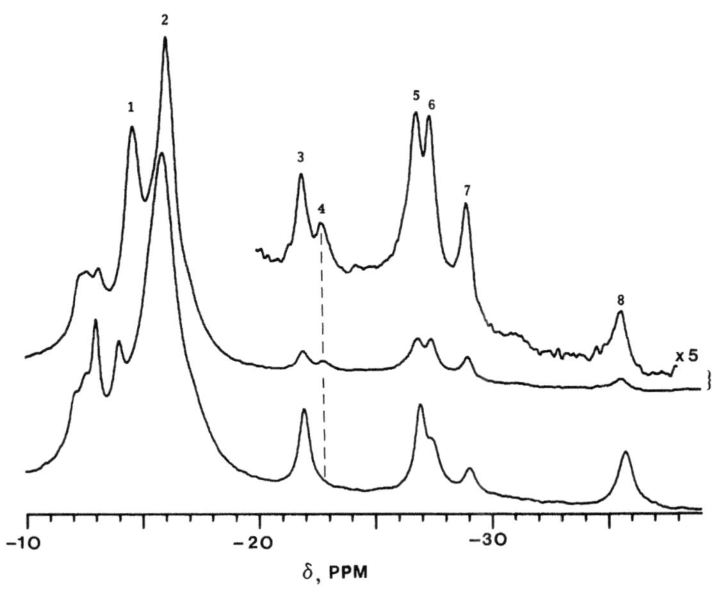

Fig. 5. Observation of NDP in hypoxic, but not normoxic, maize root tips by ^{31}P-NMR. The upper spectrum is from root tips perfused with nitrogen-saturated 50 mM glucose. High levels of cytoplasmic inorganic phosphate are apparent (peak *1*), and it is closer to the vacuolar inorganic phosphate signal (peak *2*), indicating cytoplasmic acidosis, compared to the lower spectrum. A peak attributed to NDP (*4*) is clearly visible in the upper spectrum. However, in root tips perfused with oxygenated 30 mM succinate, no ADP signal is apparent (lower spectrum). Assignments for other peaks: *3* γNTP; *5* αNTP and NDP; *6*, *7* uridine diphosphoglucose; *8* βNTP

α-phosphates of NDP and NTP appear to be impossible to resolve adequately even at the highest field strengths currently used. Hence, with high-frequency spectrometers, the concentration of NDP in plant cells can be determined from the intensity of the βNDP resonance, using the procedures outlined above. And the concentration of NTP can be determined from the intensities of both the γNTP and βNTP lines.

At low magnetic field strengths the concentration of NTP can be determined from the intensity of the βNTP line, and the concentration of NDP is given by subtracting the intensity of the βNTP line from the intensity of the γNTP + βNDP (these not being resolved) (e.g., Freeman et al. 1983). However, these measurements are more prone to error than measurements made at higher field. This is because the βNTP resonance in vivo is broader than that of the γNTP resonance, making its integration less accurate unless spectra are of very high signal to noise.

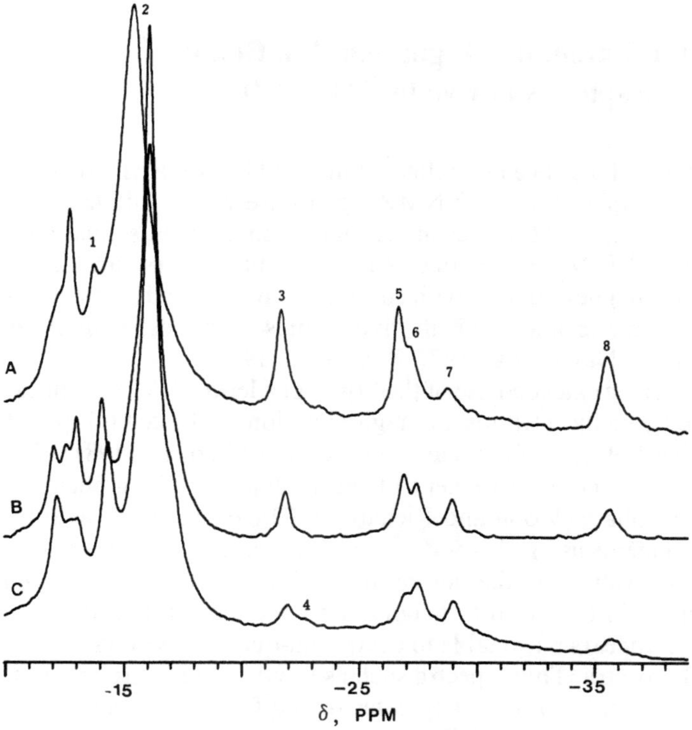

Fig. 6. Changes in the concentration of NTP, and the free energy for hydrolysis of ATP, observed by ^{31}P-NMR. Exposure of maize root tips to oxygenated 30 mM succinate (*A*) results in highs NTP levels (peaks *3*, *5*, *8*), compared to root tips that have been provided with no external respiratory substrate (*B*). Inhibition of oxidative phosphorylation by 1 mM salicylhydroxamic acid plus 0.1 mM KCN leads to a further reduction in NTP levels. Comparison of *B* and *C* also indicates that the Gibbs' free energy for hydrolysis is lower in *C*, for the products of ATP hydrolysis – cytoplasmic inorganic phosphate (peak *1*), ADP (peak *4*) and H^+ (increase indicated by rightward shift of peak *1*) – all increase in concentration as the concentration of ATP apparently falls

In normoxic maize root tips we have been unable to detect a ^{31}P-NMR signal attributable to NDP (Roberts et al. 1985), our results indicating that the ratio NTP/NDP exceeds 25. Only when the production of ATP is inhibited can we observe signals from NDP. Figures 5 and 6 illustrate this point. Improvements in the sensitivity of NMR spectrometers should allow us to observe free NDP over the entire range of physiological conditions.

^{31}P-NMR signals from NMP's are difficult to observe for one of two reasons. First, the total NMP concentration observed in perchloric acid extracts of normoxic tissue is far lower than those for NDP and NTP (Saglio et al. 1980) – too low to be detectable by ^{31}P-NMR. Second, when the concentration of NMP increases, as in hypoxic tissue (Saglio et al. 1980), the signal becomes confused among ^{31}P signals from other monophosphate esters (e.g., Figs. 3 and 5). Thus, ^{31}P-NMR cannot be used to determine the popular quantity known as the energy charge (Atkinson 1977).

6 Estimation of the Extent of Magnesium Ion Binding to Nucleoside Triphosphates in vivo by ^{31}P-NMR

The binding of ions by ATP has been investigated in detail by NMR. The chemical shifts of nucleoside triphosphate ^{31}P-NMR signals are very sensitive to the presence of magnesium ion, the βNTP resonance in particular. This is illustrated in Fig. 7. Comparison of ^{31}P-NMR spectra of ATP, in the presence of various relative concentrations of magnesium ion, with spectra of nucleotide triphosphates in vivo (e.g., Fig. 7) suggests that NTP in vivo is present almost entirely as MgNTP (Burt et al. 1976; Navon et al. 1977; Wray et al. 1983).

Other investigators have extended use of the Mg-dependence of ATP chemical shifts to estimate the free concentration of magnesium ion in the cell (Gupta et al. 1978; Gupta and Yushok 1980; Gupta and Moore 1980; Shao et al. 1981). The approach is analogous to the measurement of intracellular pH from chemical shifts of inorganic phosphate (Moon and Richards 1973). However, as with intracellular pH measurements using ^{31}P-NMR, there is potential for error because of uncertainties as to the validity of the calibration curves, obtained in vitro, used to relate NTP chemical shifts in vivo to magnesium ion concentration (Gadian 1982). Nevertheless, it appears reasonable to draw semi-quantitative conclusions from these ^{31}P-NMR spectra. Thus, spectra such as those in Fig. 7 strongly suggest that there is not a significant excess (greater than a few tens of percent) of free magnesium ion over nucleoside triphosphates in the cytoplasms of most eukaryotic cells. This conclusion is based on the observation that the chemical shift of the αATP ^{31}P resonance in vivo, and its line width relative to the line width of the βATP resonance, is consistent with a ratio of ATP : Mg^{2+} of approximately 1 (Fig. 7). Since the tissue concentration of ATP in glucose-fed maize root tips is approximately 0.5–0.6 mM (Roberts et al. 1984 b), one can estimate the concentration of Mg-ATP + Mg^{2+} in the cytoplasm to be approximately 1 mM. This is an important piece of information, since innumerable biochemical reactions are

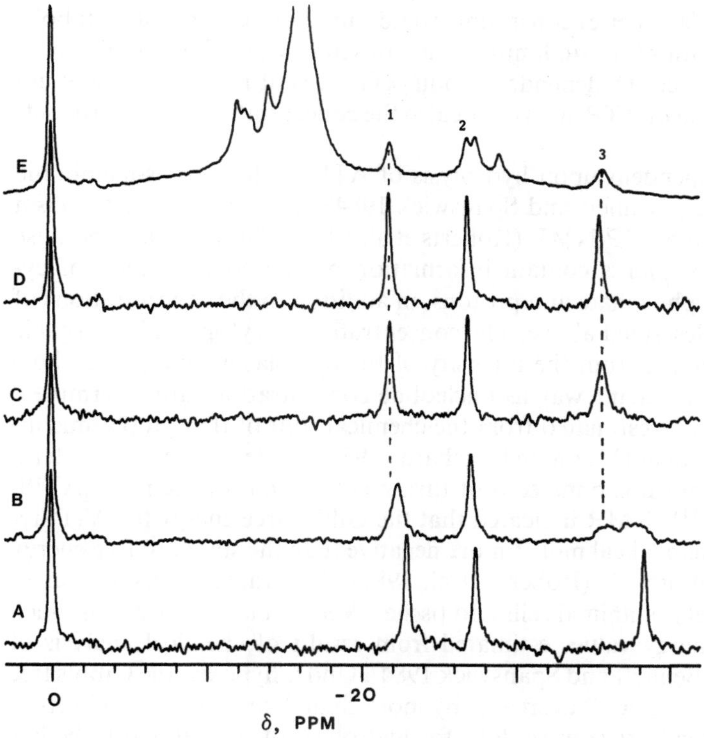

Fig. 7. Effect of magnesium ion on the chemical shifts and linewidths of the ^{31}P-NMR signals from ATP. Spectra *A–D* are of 2 mM NaATP, in 50 mM KCl plus 10 mM Tris (pH 7.5 with MES). *A* no magnesium ion present; *B* 1 mM $MgCl_2$ added; *C* 2 mM $MgCl_2$ added; *D* 3 mM $MgCl_2$ added. Spectrum *E* is of oxygenate maize root tips; peak assignments; *1* γNTP; *2* αNTP; *3* βNTP

found to be very sensitive to the concentration of magnesium ion in vivo – a result which raised the possibility that changes in the concentration of free magnesium ion might be a significant factor controlling certain metabolic processes in vivo.

7 Measurement of the Gibbs' Free Energy for Hydrolysis of ATP

The presence of ATP in cells is only one pre-requisite for metabolism in vivo. In addition, the ATP must be capable of performing work as it is used by enzymes. These two requirements must be met before any kinetic controls of metabolism can come into play.

As is well known, net synthesis of a product of a reaction can only occur when the ratio of reactants to products exceeds the equilibrium ratio. Also, as this ratio approaches the equilibrium value, the rate at which products are synthesized can decrease.

Thus for kinetic (allosteric) constraints to dominate the control of metabolic fluxes, mass action ratios for rate-limiting reactions must be far from equilibrium. Consequently, for reactions dependent upon ATP, it is of interest to know not only the concentration of ATP in vivo, but also the concentrations of the products of ATP utilization.

For reactions dependent upon hydrolysis of ATP, such as trans-membrane proton pumping (e.g., Bennett and Spanswick 1984), this aspect of metabolism can be studied in vivo by ^{31}P-NMR (Roberts et al. 1985). This is possible because ^{31}P-NMR spectra of plants contain information on the concentration of cytoplasmic inorganic phosphate and protons, in addition to the concentrations of ATP and ADP, as described above. The concentration of cytoplasmic inorganic phosphate is determined from the intensity of the cytoplasmic inorganic phosphate ^{31}P signal, in the same way as nucleotide concentrations are determined. Cytoplasmic pH can be estimated from the chemical shift of the cytoplasmic inorganic phosphate signal (Moon and Richards 1973), as described in Chapter 5 of this Volume. In normoxic maize root tips, measurement of the ratio [ATP]/[ADP] [P_i] [H^+] by ^{31}P-NMR indicated that the Gibbs' free energy for ATP hydrolysis was more than 6 kcal mol^{-1} more negative than the standard free energy (i.e., $\Delta G \sim -14$ kcal mol^{-1}) (Roberts et al. 1985). This value is very similar to that found in a variety of animal cells and tissues (Veech et al. 1979), and is much greater than free energy values estimated from results of chemical analysis of plant extracts (e.g., Bennett and Spanswick 1984). Under hypoxia, the Gibb's free energy for hydrolysis of ATP decreases by more than 2 kcal mol^{-1} (Roberts et al. 1985). Only in dead tissue were ΔG's for hydrolysis of ATP found to be less than ~ -10.5 kcal mol^{-1} (Roberts et al. unpublished observations). It is not clear whether this last result indicates that coordinated metabolism cannot be sustained at a ΔG_{ATP} less than ~ -10 kcal mol^{-1}, or that some other factor caused death. However, this observation certainly raises the question "What is the minimum ΔG_{ATP} necessary to sustain metabolism in vivo?"

A potential limitation of the ^{31}P-NMR method for measurement of ΔG_{ATP} arises from the fact that the method does not permit distinction of the various nucleotides (ATP, CTP, GTP etc.). Hence, errors in the measurement of ΔG_{ATP} can occur if the ratio [NTP]/[NDP] is not equal to [ATP]/[ADP]. In fact, such errors are most probably small, because the concentration of adenine nucleotides usually is much greater that the concentrations of other nucleotides, particularly in nongrowing tissues (R. H. Nieman, personal communication). If nonadenine nucleotides do comprise a significant proportion of the nucleotide pool, it is important to evaluate the possibility of differences in the precise values of [NTP]/[NDP] for the different nucleotides. We have found that the ratio [ATP]/[ADP] is always greater than or equal to the equivalent ratio for other nucleotides (Roberts et al. 1985). This result is not surprising, for nucleoside triphosphates other than ATP are largely synthesized at the expense of ATP due to action of nucleoside diphosphokinase: ATP + NDP = ADP + NTP, the equilibrium constant for this reaction being ~ 1 (Reich and Sel'kov 1981).

Figure 8 illustrates a situation in which the free energy for hydrolysis of ATP changes without any change in concentration of NTP. Thus, while changes in ATP levels or energy charge do not always correlate with metabolic activity (e.g.,

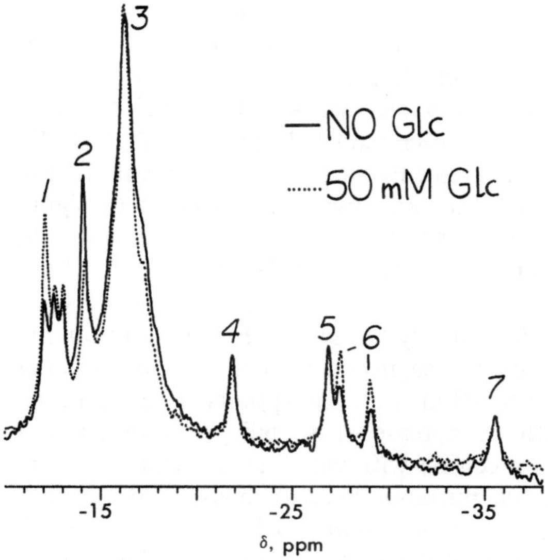

Fig. 8. Effect of external glucose on the concentration of NTP and the free energy of hydrolysis of ATP in maize root tips. Spectra are of the same normoxic maize root tip sample, either before or after exposure to 50 mM glucose. The glucose stimulates respiration ca. 50% (Roberts et al. 1984b), without affecting NTP levels (peaks *4*, *5*, and *7*). This was previously reported by Saglio and Pradet (1980). The large decrease in the concentration of cytoplasmic inorganic phosphate (peak *2*) in response to glucose – that occurs as levels of glucose-6-phosphate (peak *1*) and uridine diphosphoglucose (peaks *6*) increase – indicates that the free energy for hydrolysis of ATP has increased. (Roberts et al. 1984b)

the rate of ATP synthesis) (Saglio and Pradet 1980), changes in ΔG_{ATP} do, such that a decrease in the availability of ATP (inhibition of mitochondrial activity) results in a decrease in the amount of useful work ATP can do (per mol) (Roberts et al. 1985).

8 Relationship of in vivo ^{31}P-NMR Measurements to What Is Seen by Other Methods

Accuracy. Here one is concerned with the relationship between what is measured by an experimental method, and the true in vivo value of the particular parameter whose measurement is attempted. The attractive feature of the ^{31}P-NMR method is that measurements of levels of free nucleotides, etc. are made without disturbing the plant system in any way. Provided the nature of the chemical groups that contribute to the various peaks in NMR spectra are correctly identified, no artifacts can destroy or distort the direct and simple relationship between what is measured by NMR and what exists in vivo. Further, ^{31}P-NMR permits observation of only freely mobile metabolites, i.e., only those metabolites that are free to participate in metabolism – either as reactants or as allosteric effectors (provided they are present at a free concentration greater than ca. 0.1 mM). These characteristics of the ^{31}P-NMR method contrast with the characteristics of methods that rely on chemical analysis of plant extracts. In chemical methods uncertainties arise from problems of breakdown of metabolite such as ATP, which accounts in part for the disparity in the literature between reported values of [ATP]/[ADP] (Atkinson 1977; Pradet and Raymond 1983). Extraction and chemical analysis also does not permit distinction between "bound" and "free" metabolites; significant amounts of inorganic phosphate and ADP appear to occur in an immobile physical state in some animal tissues (e.g., Freeman et al. 1983).

Precision. NMR spectroscopy is an intrinsically less sensitive method for quantitation than are most chemical methods. Hence, in general, concentrations can be determined with less precision by the NMR method. Two points diminish the impact of this disadvantage. First, with the increased availability of more sensitive high-field NMR spectrometers, the precision with which measurements can be made is sufficient for most physiological purposes. Second, in vivo NMR permits multiple, sequential measurements over many hours to be made on a single sample. This not only permits averaging of separate measurements, but also enables variability from sample to sample to be reduced: "before" and "after" measurements are made on the same sample.

Compartmentation (Where are the metabolites visible by ^{31}P-NMR located in the cell?). Conventional methods for analysis of metabolites in tissues do not permit determination of concentrations in different intracellular compartments. This is a significant limitation, for the concentrations of metabolites such as ATP and ADP differ greatly between compartments such as cytoplasm and mitochondria (Klingenberg and Heldt 1982; Stitt et al. 1982). ^{31}P-NMR is capable of distinguishing metabolites in particular intracellular compartments under two types of circumstance. The first type is found when the environments in the various compartments differ sufficiently for metabolites in each compartment to give NMR signals of different chemical shift. For example, the large difference in pH between vacuoles and cytoplasm permits resolution of cytoplasmic and vacuolar inorganic phosphate ^{31}P-NMR signals (Roberts et al. 1980). In contrast, only one nucleotide signal is observed. The second type of circumstance that allows metabolites in only one compartment to be observed by NMR is found when the freedom of metabolites to move (tumble, diffuse) differs from compartment to compartment. For example, serotonin and ATP in blood platelet granules is relatively immobile such that it cannot be seen by NMR (Ugurbil et al. 1979; Costa et al. 1979); only cytoplasmic ATP in human blood platelets is sufficiently mobile to be observed by ^{31}P-NMR. It would appear that the same is true for the nucleotides in plants. The principle reasons for this view are (1) ATP/ADP ratios observed by ^{31}P-NMR are much closer to values observed for cytoplasm in cell fractionation studies (Stitt et al. 1982), being far higher than the ratios observed in mitochondria. Similarly, the free energy for hydrolysis of ATP determined by NMR (Roberts et al. 1985) corresponds closely to that determined for the cytoplasm – not the mitochondria – in a variety of animal cells (Veech et al. 1979). (2) In some animal tissues (which, in general, contain a far greater volume of mitochondria), comparisons between ^{31}P-NMR and chemical analysis have shown that much of the ADP and some of the inorganic phosphate measured in analyses of tissue extracts is not detectable in vivo by ^{31}P-NMR. This indicates that there are significant amounts of immobilized ADP and inorganic phosphate in vivo. It is possible that much of these phosphates are located in the mitochondria: even animal tissues that have mitochondria occupying as much as 40% of the cell volume have not given ^{31}P-NMR signals attributable to mitochondria (see discussion in Chap. 5, this Vol.).

9 Conclusions

^{31}P-NMR is a useful technique for quantitating the energy status in plant tissues. Tissue concentrations of nucleotides can be determined in vivo, without the uncertainties of breakdown and/or interferences often associated with methods involving assay of plant extracts. The ability of the ^{31}P-NMR method to measure concentrations of cytoplasmic inorganic phosphate and H^+, as well as nucleotides, represents the only means of directly determining the Gibbs' free energy for hydrolysis of ATP in plant cells. It appears possible that these attributes, together with the use of ^{31}P-NMR to measure the rate of ATP synthesis in the steady state, in vivo (Roberts et al. 1984b), might help attempts to understand energy metabolism.

References

Atkinson DE (1977) Cellular energy metabolism and its regulation. Academic Press, New York
Becker ED (1980) High resolution NMR. Theory and chemical applications. 2nd edn. Academic Press, New York
Bennet AB, Spanswick RM (1984) H^+-ATPase activity from storage tissues of Beta vulgaris. II. H/ATP stoichiometry of an anion-sensitive H^+-ATPase. Plant Physiol 74:545–548
Burt CT, Glonek T, Barany M (1976) Analysis of phosphate metabolites, the intracellular pH, and the state of adenosine triphosphate in intact muscle by phosphorus nuclear magnetic resonance. J Biol Chem 251:2584–2591
Costa JL, Dobson CM, Kirk KL, Poulsen FM, Valeri CR, Vecchione JJ (1979) Studies of human platelets by ^{19}F and ^{31}P NMR. FEBS Lett 99:141–146
Dawson MJ, Gadian DG, Wilkie DR (1977) Contraction and recovery of living muscles studied by ^{31}P nuclear magnetic resonance. J Physiol 267:703–735
Erecinska M, Wilson DF (1982) Regulation of cellular energy metabolism. J Membr Biol 70:1–14
Farrar TC, Becker ED (1971) Pulse and fourier transform NMR: Introduction to theory and methods. Academic Press, New York
Freeman D, Bartlett S, Radda G, Ross B (1983) Energetics of sodium transport in the kidney. Saturation transfer ^{31}P-NMR. Biochim Biophys Acta 762:325–336

Gadian DG (1982) Nuclear magnetic resonance and its application to living systems. Oxford Univ Press, Oxford
Gadian DG, Radda GK (1981) NMR studies of tissue metabolism. Annu Rev Biochem 50:69–83
Gupta RK, Moore RD (1980) ^{31}P NMR studies of intracellular free Mg^{2+} in intact frog skeletal muscle. J Biol Chem 255:3987–3993
Gupta RK, Yushok WD (1980) Noninvasive ^{31}P NMR probes of free Mg^{2+}, MgATP, and MgADP in intact Ehrlich ascites tumor cells. Proc Natl Acad Sci USA 77:2487–2491
Gupta RK, Benovic JL, Rose ZB (1978) The determination of the free magnesium level in the human red blood cell by ^{31}P-NMR. J Biol Chem 253:6172–6176
Klingenberg K, Heldt HW (1982) The ADP/ATP translocator in mitochondria and its role in intracellular compartmentation. In: Sies H (ed) Metabolic compartmentation. Academic Press, New York
Lipmann F (1941) Metabolic generation and utilization of phosphate bond energy. Adv Enzymol 1:99–162
Martin J-B, Bligny R, Rebeille F, Douce R, Legauy J-J, Matheiu Y, Guern J (1982) A ^{31}P nuclear magnetic resonance study of intracellular pH of plant cells cultivated in liquid medium. Plant Physiol 70:1156–1161
Moon RB, Richards JB (1973) Determination of intracellular pH by ^{31}P magnetic resonance. J Biol Chem 249:7276–7278
Navon G, Ogawa S, Shulman RG, Yamane T (1977) ^{31}P nuclear magnetic resonance studies of Ehrlich ascites tumor cells. Proc Natl Acad Sci USA 74:87–91
Pradet A, Raymond P (1983) Adenine nucleotide ratios and adenylate energy charge in energy metabolism. Annu Rev Plant Physiol 34:199–224
Reich JG, Sel'kov EE (1981) Energy metabolism of the cell. A theoretical treatise. Academic Press, New York
Roberts JKM (1984) Study of plant metabolism in vivo using NMR spectroscopy. Annu Rev Plant Physiol 35:375–386
Roberts JKM, Jardetzky O (1981) Monitoring of cellular metabolism by NMR. Biochim Biophys Acta 639:53–76
Roberts JKM, Ray PM, Wade-Jardetzky N, Jardetzky O (1980) Estimation of cytoplasmic and vacuolar pH in higher plants by ^{31}P-NMR. Nature 283:870–872
Roberts JKM, Wemmer D, Ray PM, Jardetzky O (1982) Regulation of cytoplasmic and vacuolar pH in maize root tips under different physiological conditions. Plant Physiol 69:1344–1347
Roberts JKM, Callis J, Wemmer D, Walbot V, Jardetzky O (1984a) The mechanism of cytoplasmic pH regulation in hypoxic maize root tips and its role in survival under hypoxia. Proc Natl Acad Sci USA 81:3379–3383
Roberts JKM, Wemmer D, Jardetzky O (1984b) Measurement of mitochondrial ATPase activity in maize root tips by saturation transfer ^{31}P nuclear magnetic resonance. Plant Physiol 74:632–639
Roberts JKM, Linker CS, Benoit AG, Jardetzky O, Nieman RH (1984c) Salt stimulaton of phosphate uptake in maize root tips studied by ^{31}P nuclear magnetic resonance. Plant Physiol 75: 947–950
Roberts JKM, Lane AN, Clark RA, Nieman RH (1985) Relationships between the rate of synthesis of ATP and the concentrations of reactants and products of ATP hydrolysis in maize root tips, determined by ^{31}P nuclear magnetic resonance. Arch Biochem Biophys 240:712–722
Saglio PH, Pradet A (1980) Soluble sugars, respiration, and energy charge during aging of excised maize root tips. Plant Physiol 66:516–519
Saglio PH, Raymond P, Pradet A (1980) Metabolic activity and energy charge of excised maize root tips under anoxia. Control by soluble sugars. Plant Physiol 66:1053–1057
Scott AI, Baxter RL (1981) Applications of ^{13}C NMR to metabolic studies. Annu Rev Biophys Bioeng 8:1–25
Shao TW, Pieper GM, Salhany JM, Eliot RS (1981) Measurement of free magnesium in perfused and ischemic arrested heart muscle. A quantitative phosphorus-31 nuclear magnetic resonance and multiequilibria analysis. Biochemistry 20:1399–1403

Shaw D (1976) Fourier transform NMR spectroscopy. Elsevier, Amsterdam
Shulman RG, Brown TR, Ugurbil K, Ogawa S, Cohen SM, den Hollander JA (1979) Science 205:160–166
Stitt M, Lilley R McC, Heldt HW (1982) Adenine nucleotide levels in the cytosol, chloroplasts and mitochondria of wheat leaf protoplasts. Plant Physiol 70:971–977
Ugurbil K, Holmsen H, Shulman RG (1979) Adenine nucleotide storage and secretion in platelets as studied by ^{31}P nuclear magnetic resonance. Proc Natl Acad Acid Sci USA 76:2227–2231
Veech RL, Lawson JWR, Cornell NW, Krebs HA (1979) Cytosolic phosphorylation potential. J Biol Chem 254:6538–6547
Wilson DF, Stubbs M, Veech RL, Erecinska M, Krebs HA (1974) Equilibrium reactions between the oxidation-reduction reaction and the adenosine triphosphate synthesis in suspensions of isolated liver cells. Biochem J 140:57–64
Wray V, Schiel O, Berlin J (1983) High field phosphorus-31 nuclear magnetic resonance investigation of the phosphate metabolites in cell suspension cultures of *Nicotiana tabacum*. Z Pflanzenphysiol 112:215–220

The Use of NMR Spectroscopy to Follow Deuterium in Studies of Fungal Metabolism

C. Abell

1 Introduction

The biosynthesis of secondary metabolites by fungi continues to be one of the most studied areas in bioorganic chemistry. This is in part a result of three contributory factors, foremost amongst which is the diverse and fascinating chemistry involved. Fungi produce numerous secondary metabolites, many of which are either polyketide or terpenoid in origin (Turner and Aldridge 1983; Turner 1971). These natural products are produced during a distinct phase of the life cycle of the organism and often have a limited taxonomic distribution. The second factor is the relative accessibility of the subject. Fungi can be conveniently grown as surface or shake cultures in the laboratory. Each experiment only takes a few weeks at most and is not seasonally dependent. Finally, and most relevant in the current context is the complementarity between fungal experiments and the use of high field NMR spectroscopy, which makes this technique ideal for investigating biosynthetic pathways in these systems. This complementarity is in terms of the kind of molecule (molecular weight up to 1000), the yield of metabolite from a feeding experiment (often 0.01–0.1 mmol) and the level of incorporation (usually in the range 0.5–5%). It is therefore not surprising that in the years since the introduction of ^{13}C and ^{2}H-NMR spectroscopy to biosynthetic studies, parallel strides have been made in the understanding of biosynthetic pathways and the development of the NMR techniques.

In this review, the use of NMR spectroscopy to monitor deuterium in fungal metabolism is described. The literature is covered up to the end of 1983 in an exemplary rather than comprehensive manner. One theme which is developed is the advantage of using a label for hydrogen to probe fine details of a biosynthetic pathway. The hydrogen distribution is sensitive to many processes, including oxidation, hydride shifts and aromatisation, which do not perturb the carbon distribution. Indeed, it is this distinction which makes the combined use of deuterium and ^{13}C labels such a powerful technique.

Being safe to handle, inexpensive and available at a high level of enrichment, deuterium is an extremely useful isotope of hydrogen for biosynthetic studies, and has been widely used since its first application in 1936 (Erlenmeyer et al. 1936). Until as recently as 1974 the best method for analysing deuterium incorporation was by mass spectroscopy, but this technique has been largely superceded by the

Abbreviations. $Cr(acac)_3$, tris[acetylacetonato]chromium(III); $Eu(hfbc)_3$, tris[3-heptafluorobutanoyl-(−)-camphorato]europium(III); $Eu(fod)_3$, tris[6,6,7,7,8,8,8-heptafluoro-2,2-dimethyl-3,5-octanedionato]europium(III); FID, free induction decay; NMR, nuclear magnetic resonance; NOE, nuclear Overhauser effect.

introduction of nuclear magnetic resonance spectroscopy. The detection of deuterium directly by ^{2}H-NMR spectroscopy or indirectly by isotopic shifts and heteronuclear coupling in the ^{13}C-NMR spectrum provide rapid non-destructive methods for identifying the sites of enrichment in a molecule. These two techniques are described separately in Sections 2 and 3 of this chapter.

2 Deuterium NMR Spectroscopy

2.1 Background and Techniques

2.1.1 The Nuclear Magnetic Properties of Deuterium

The nuclear magnetic properties of deuterium are summarised and compared with those of protium in Table 1. Two of these properties are particularly significant in determining the spectroscopic behaviour, deuterium is a quadrupole nucleus with a nuclear spin of one and has a low gyromagnetic ratio.

The resonance frequency and spectral dispersion are proportional to the gyromagnetic ratio, and so are smaller for deuterium than protium by the factor $\gamma_D/\gamma_H = 0.154$. Using a 9.38 Tesla superconducting magnet, the deuterium resonance frequency is 61.42 MHz with 61.42 Hz per ppm compared with 400 MHz and 400 Hz respectively for protium. The low gyromagnetic ratio means that deuterium is not very sensitive (the sensitivity of a nucleus varies as γ^3). However, as sensitivity is also a function of field strength, by operating at high field it is possible to detect deuterium easily at natural abundance (0.0156%) with as little as 0.1 mmol of material (MW = 500) in a few hours.

The relaxation behaviour of deuterium is dominated by very efficient intramolecular quadrupole relaxation. Under conditions of extreme narrowing the quadrupolar spin-lattice relaxation time T_1 and the spin-spin relaxation time T_2 are identical and in the range 20 ms to 20 s. The very short lifetimes of excited spin states can cause serious broadening of the NMR signal.

2.1.2 The Deuterium NMR Spectrum

An NMR spectrum is characterised in terms of the chemical shift, coupling, linewidth and intensity of the resonances. The chemical shifts of deuterons correlate

Table 1. The magnetic properties of deuterium and protium

	^{2}H	^{1}H
Nuclear moment	0.86	2.79
Nuclear spin	1	0.5
Quadrupole moment	0.0027	–
Sensitivity	0.00965	1.00
Relative gyromagnetic ratio	0.154	1.00
Resonant frequency (MHz)	61.42	400.0

closely with those for the corresponding protons because the shielding of a nucleus is largely determined by the local electronic environment of that nucleus which is not greatly altered by isotopic substitution. The small differences in chemical shift which do occur due to isotopic shifts are less than 0.05 ppm and are often undetected unless a mixture of deuteriated species is present e.g. $-CD_3$, $-CD_2H$, $-CDH_2$. In this example the substitution of each additional proton moves the deuterium resonance downfield. The correspondence in chemical shift between protium and deuterium means that assignments can be made directly from the ^{1}H-NMR spectrum.

The magnitude of the coupling constant between two nuclei is proportional to the product of their gyromagnetic ratios. Consequently, the heteronuclear coupling of deuterium to protium and carbon is 6.51 times smaller than the corresponding coupling of protium. Deuterium spectra are routinely recorded under conditions of broadband proton decoupling, as this can sharpen up somewhat the otherwise broad resonances. Homonuclear deuterium-deuterium coupling is very small and is not detected.

The most striking feature of a deuterium spectrum is often the broadness of the resonances and the lack of signal resolution. The broad lines are a direct result of the rapid quadrupole relaxation of the deuterium nucleus. The linewidth of a resonance due to a deuteron attached to a rigid molecule varies as a function of molecular size as shown in the equation:

$$W \propto \frac{4\pi\eta a^3}{3kT},$$

where W is the linewidth at half height, η the viscosity of the solvent, T is the absolute temperature and a the molecular size (Glasel et al. 1973). The great sensitivity of linewidth to increasing molecular size severely limits the application of ^{2}H-NMR spectroscopy in biosynthetic studies. This was clearly shown when the use of the technique was considered for studying vitamin B_{12} biosynthesis. The ^{2}H-NMR spectrum of heptamethyl dicyanocobyrinate (MW over 1000) deuteriated on the corrin nucleus consisted only of very broad resonances ($W = 35$–40 Hz) with correspondingly low sensitivity (Battersby et al. 1982).

Intermolecular association can effectively increase the size and rotational correlation time of the deuteriated species. Dramatic increases in linewidth from 12 to 108 Hz were observed on increasing the concentration of a solution of 8-deuteroadenosine from 0.01 to 1.0 M (Egan 1976). Biosynthetic experiments are potentially very prone to such effects, as it is often desirable to run spectra of very concentrated samples to improve signal-to-noise. This was the case in a study of nocardicin A biosynthesis. The ^{2}H-NMR spectrum of a 0.12 M solution of nocardicin A (**1**) derived from (2S, 3R)–[3-^{2}H]-serine includes one major resonance corresponding to deuterium at C-4 (Fig. 1). The linewidth of this signal was reduced from 30 to 15 Hz by recording the spectrum at 45 °C instead of 25 °C. Under these conditions it was possible to assign the resonance to deuterium in the 4-α position and so show that the formation of the β-lactam ring occurs with inversion at C-4 (Townsend et al. 1983).

Signal intensities in ^{2}H NMR spectra can be accurately integrated. The short T_1 relaxation times and absence of nuclear Overhauser enhancement, both con-

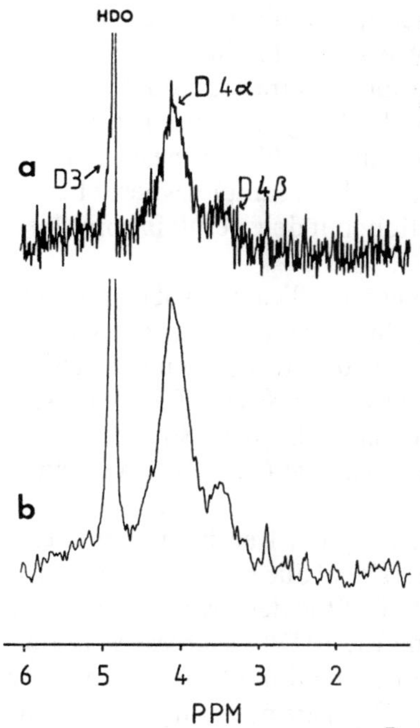

Fig. 1 a, b. ^{2}H-NMR spectra of nocardicin A (1) after incorporation of (2 S,3 R)-[3-^{2}H]serine. **a** unenhanced, **b** sensitivity enhanced. (Townsend et al. 1983)

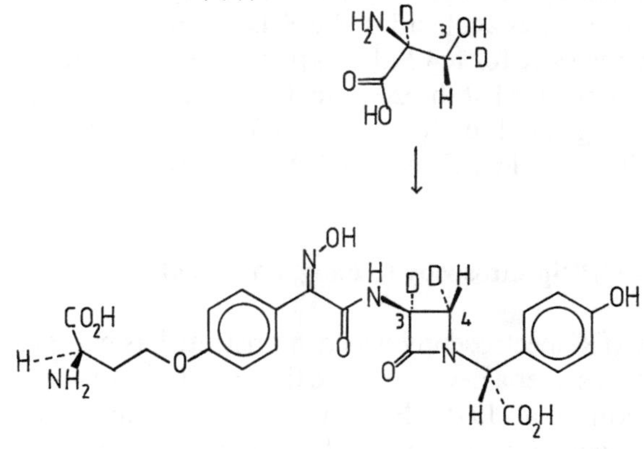

(1)

sequences of the quadrupolar relaxation, reduce the risk of partial saturation and ensure that the observed signal intensity is proportional to the extent of deuteriation. It is therefore possible to acquire spectra using rapid pulsing without delays.

From the above description of the characteristic features of deuterium spectra, it is clear that the nuclear properties of the deuteron restrict greatly the utility of ^{2}H-NMR. The major problems stems from the poor spectral dispersion and the inherent broadness of deuterium resonances.

2.1.3 Recording a Deuterium NMR Spectrum

The deuteriated sample obtained from the biosynthetic experiment (1–50 mg) is made up as a solution in the appropriate protonated solvent. Where possible, low viscosity solvents such as dicloromethane or acetone should be used. In cases where there is extremely low incorporation, solvents depleted in deuterium can be used. On high-field NMR spectrometers deuterium spectra are recorded while locked on the fluorine signal of a fluorinated standard, e.g., hexafluorobenzene. The fluorinated standard is used either as a co-solvent (5–10%) or neat in an insert in the NMR tube. The latter method is useful for those samples where chemical shift proves to be sensitive to solvent composition, but does result in some loss of resolution.

To record ^{2}H-NMR spectra, it is necessary to use a dedicated probe or a multinuclear probe tuned to deuterium, ideally with a fluorine lock and the capability to decouple proton. Decoupled spectra are recorded with continuous proton broadband decoupling using 2–3 W of decoupling power or less if a composite decoupling modulation such as Waltz is used (Shaha et al. 1983). Spectra are recorded over the same ppm range as for proton. Typically at 61.4 MHz, a spectral width of 800 Hz and a 90 ° pulse are used.

The acquisition delay is determined by watching the free induction decay (FID) decay until it has died away completely. The signal due to deuterium at natural abundance in the solvent can dominate the FID if the deuterium enrichment in the sample is low. In order to avoid truncating the FID in such cases, a long acquisition time (5 s) is used to allow for the slow relaxation of the solvent. For rigorously quantitative work, long delays (at least $5T_1$S) are necessary. The acquisition time determines the block size. After acquisition the FID is zero-filled once by transfer to a large block of zeros (8 K for a 4 K FID) prior to Fourier transformation. Signal-to-noise can be improved at the expense of resolution by applying a matched filter (line broadening equal to T_2*) to the FID. This is well illustrated in the spectra of nocardicin A in Fig. 1 (Townsend et al. 1983).

2.2 Applications of Deuterium NMR Spectroscopy to Fungal Metabolism

Although the potential of ^{2}H-NMR spectroscopy was demonstrated as early as 1964 (Diehl and Leipert 1964), it was over a decade before the technique was used to study fungal metabolism (Bycroft et al. 1975). Encouraged by that pioneering experiment and the increased availability of high-field NMR spectrometers, bioorganic chemists have rapidly developed ^{2}H-NMR spectroscopy into a powerful tool to serve their needs (Garson and Staunton 1979; Hutchinson 1982). The many experiments in which direct observation of deuterium by NMR spectroscopy has been used as the main analytical method can be classified according to whether the deuterons label a hydrogen, a molecule or a bond. These distinctions are further refined in the following discussion.

2.2.1 Administering the Precursor

In any feeding experiment, labelled precursors are diluted by unlabelled endogenous material to an extent which depends on the type of organism and the timing of the feeding. The low natural abundance of deuterium (0.0156%) and the high level of enrichment to which precursors can be labelled (100%) mean that even after a 6000-fold dilution of labelled precursor there will be a doubling of peak intensity over the natural abundance signal in the ^{2}H-NMR spectrum of the metabolite. In practice the lower limit for quantitative detection of deuterium incorporation is 0.1%, corresponding to a dilution of 1000 (Hutchinson et al. 1979). Incorporations of 0.5–5% are frequently obtained in fungal experiments, well within the sensitivity range of ^{2}H-NMR spectroscopy. This is not true of incorporation experiments with plants, where dilution factors of 10^6 are commonly encountered. There are nevertheless several notable examples of the use of ^{2}H-NMR spectroscopy to study secondary metabolism in plants (Rana and Robins 1983; Damtoft 1981; Dewick and Ward 1977).

The correct time to administer a labelled precursor to a fungus to obtain optimum incorporation into secondary metabolites is around the beginning of the idiophase. This is the stage in the lifecycle when the growth of the fungus has slowed down and the enzymes of secondary metabolism are becoming active. In most biosynthetic studies the timing of a feeding is not optimised once satisfactory incorporation results have been found using radiolabelled substrates. The feeding strategy used depends to some extent on the nature of the precursor and the kind of information being sought. Either the precursor can be fed in one shot or pulse-fed over several days. The latter approach maintains a fairly steady level of labelled precursor in the growing broth, and often leads to better incorporations. This is usually the best regime to use when feeding advanced precursors or singly labelled simple precursors. When feeding doubly labelled precursors, especially acetate, the extra precaution of diluting the labelled material prior to feeding reduces the complications that can arise from incorporation of adjacent labelled units. A single feeding of an undiluted simple precursor can result in multiple incorporations in the same molecule. This has been used to show connectivities between carbon atoms in different subunits of a molecule (Bradshaw et al. 1982d) but is less useful in deuterium labelling studies.

2.2.2 Establishing the Retention or Loss of a Hydrogen

This is conceptually the simplest use of a hydrogen label. By determining whether a hydrogen is retained or lost from a certain position in a precursor, the range of structures of later precursors can be narrowed or some mechanisms precluded. This was the aim of the first biosynthetic application of ^{2}H-NMR spectroscopy (Bycroft et al. 1975). L-[2-^{2}H]-cystine was incorporated into penicillin G **(2)** by

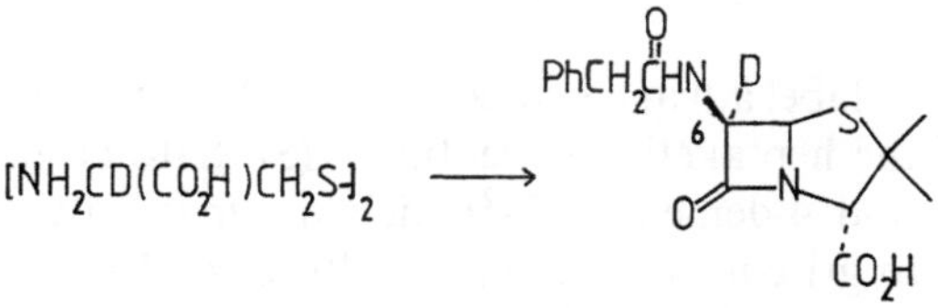

(2)

Penicillium chrysogenum. The ^{2}H-NMR spectrum of **(2)** showed that the product was deuterium-labelled specifically at H-6. Without a standard for comparison, the extent of deuterium loss from the precursor before or during the assembly of the five-membered ring could not be measured from this single experiment. Nevertheless the fact that some deuterium did remain to label H-6 indicated that an enzyme-free intermediate containing an α,β-dehydrocysteine moiety is not on the biosynthetic pathway.

A more sophisticated deuterium-labelling study was recently performed to look for hydrogen loss from the valine section of the tripeptide in the cell-free conversion of tripeptide to isopenicillin N **(3)** (Baldwin et al. 1981). (L-α-amino-δ-adipyl)-L-cysteinyl-D-[2-^{2}H,Me$_2$-^{2}H$_6$]-valine **(4)** was synthesised and converted into

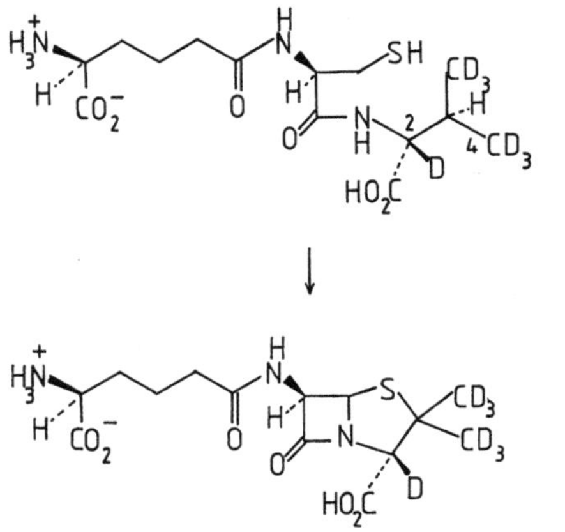

(4)

(3)

isopenicillin N. The ratio of deuterium atoms at the positions derived from C-2 and C-4,4′ of valine in the penicillin was identical to that in the starting material. Not only are no enzyme-free dehydro intermediates involved, there is no hydrogen loss at all from any of the three labelled centres.

2.2.3 Establishing Precursor-Product Relationships

Deuterium labelling can be used to label a whole molecule, identifying it from the pool of unlabelled endogenous material. The label should be positioned in the precursor so that it will not be lost in the subsequent transformations, and in such a way that its location in the product can distinguish whether or not the incorporation is specific. In these experiments the detection of deuterium from a labelled compound in the product at the expected locus is supporting evidence that that compound is an intermediate on the biosynthetic pathway. However, it should be emphasised that specific incorporation is not alone sufficient evidence to identify an intermediate as obligatory.

^{2}H-NMR spectroscopy was used to label advanced precursors in the extended investigation of the biosynthesis of the heptaketide griseofulvin **(5)** (Sato et al. 1978). [5′-^{2}H]Griseophenone B **(6)** and 4-demethyl-[5′-^{2}H]dehydrogriseofulvin **(7)** were incorporated specifically into griseofulvin by shake cultures of *Penicillium urticae*. This confirmed results from earlier studies with ^{14}C-labelled compounds and additionally furnished stereochemical information about the reduction of the 5′-6′ double bond. Deuterium from both **(6)** and **(7)** labelled only the 5′-α position of griseofulvin, showing that trans hydrogenation must occur. The same result was observed in the microbial hydrogenation of dehydrogriseofulvin

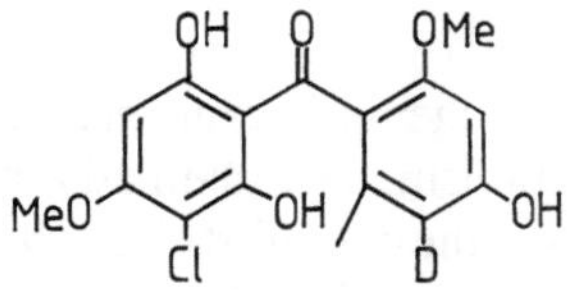

(6)

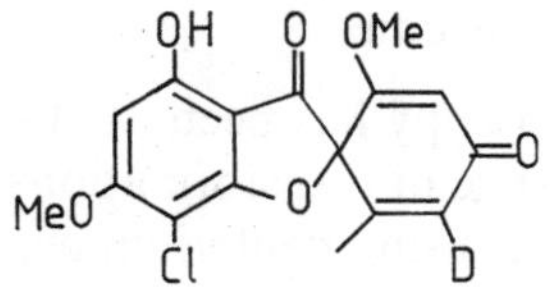

(7)

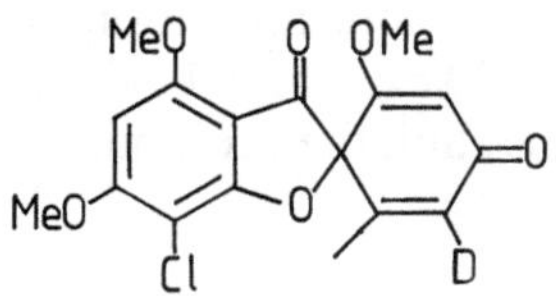

(8)

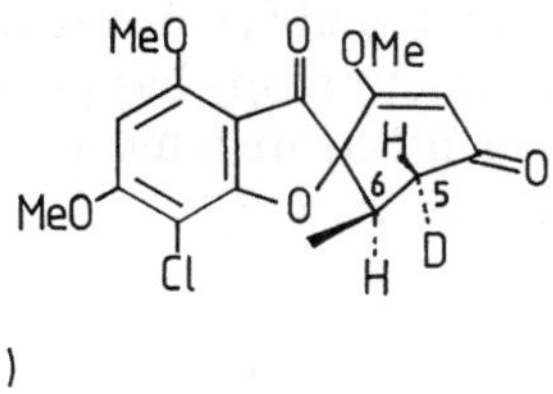

(5)

(8) by *Streptomyces cinereocrocatus*, suggesting that similar reducing enzymes may be present in both organisms (Sato et al. 1977).

The keto-aldehyde **(9)** labelled with deuterium at C-11 was incorporated specifically into citrinin **(10)** by a surface culture of *Penicillium citrinum* (Barber and

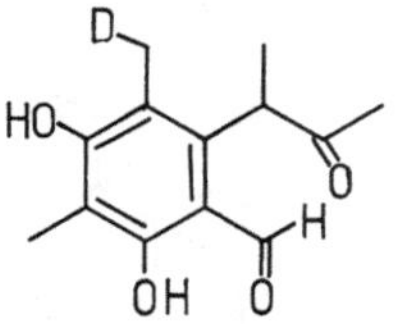

(9)

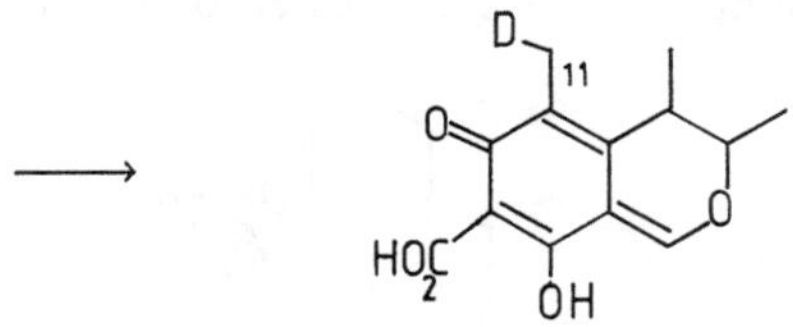

(10)

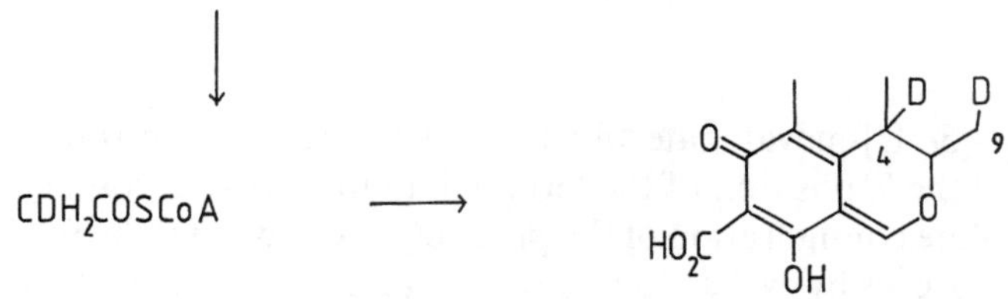

Staunton 1980a). Comparison of the intensity of the C-11 signal of labelled citrinin at δ 2.0 with the signal due to deuterium at natural abundance in the chloroform solvent in the ^{2}H-NMR spectrum indicated an enrichment of 1.6%. Less intense resonances due to deuterium at C-9 and C-4 of citrinin were also observed. The presence of these extra signals suggests an alternative fate for the ketoaldehyde, namely degradation to release monodeuterioacetyl CoA which serves as a precursor in the assembly of the polyketide chain of **(10)**. The two fates of the ketoaldehyde could not have been demonstrated as easily using other analytical techniques. The incorporation of **(9)** was dramatically improved to 80% by replacing the growing medium at the onset of citrinin production and adding ethionine, a known inhibitor of enzymic methylation (Barber and Staunton 1980b).

2.2.4 Mechanistic Studies

^{2}H-NMR spectroscopy has been used to study the stereochemistry and mechanism of a wide range of transformations in fungal metabolism. The main advantages derived from using deuterium are the ability to synthesise substrates to a high level of enrichment and the direct spectroscopic analysis of the final labelling pattern. The utility of this approach is extended considerably when substrates doubly labelled with ^{13}C and ^{2}H are used. Unfortunately the poor resolution inherent in ^{2}H-NMR spectroscopy has proved a major drawback, as the key sites which must be distinguished are often inevitably very similar and their resonances are then difficult to resolve.

2.2.4.1 Displacement Reactions

Stereospecific labelling at a methylene centre bearing a leaving group has been used to study the ring closure step in nocardicin A biosynthesis (Townsend et al. 1983) and the S_N2' displacement of allylic pyrophosphates in the biosynthesis of the fungal diterpene rosenonolactone **(11)** (Cane and Murthy 1977). In the latter study the prochiral hydrogens at C-15 of the bicyclic intermediate labda-8(17),13-dien-15-yl pyrophosphate **(12)** were labelled by incorporating mevalonates deu-

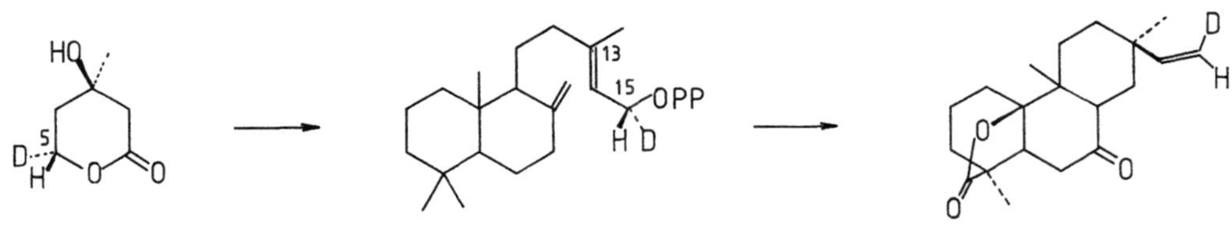

(12) (11)

teriated at C-5. Thus 5R-[5-^{2}H]-mevalonate which labels the 15-pro-R position of **(12)** was found to label the Z position of the terminal double bond in rosenonolactone. From the absolute configuration of the product, it follows that the formation of the third ring proceeds by cyclisation onto the si face of C-13. Therefore this labelling result shows that the displacement reaction takes place antarafacially. In this study it was imperative to perform the complementary feedings with 5S-[5-^{2}H]- and [5-$^2H_2$]-mevalonates because the resonances for the two terminal deuterons were extremely close ($\Delta\delta = 0.07$ ppm) and broad (3.5 and 7 Hz). Only by comparing the spectra of all three labelled species could the data be interpreted unambiguously.

2.2.4.2 Incorporation of Deuterium Atoms from the Medium

Metabolites can be universally labelled with deuterium by the simple expedient of growing the organism in a medium containing 5–10% deuterium oxide (Bardshiri and Simpson 1983). This is a useful prelude to incorporation studies with deuteriated precursors as the deuterium spectrum of the fully deuteriated sample will highlight any problems due to signal overlap and provide a reference for subsequent spectra.

Experiments to show site-specific incorporation of deuterium from the medium have been performed with whole cells and cell-free systems. The microbial hydrogenation of dehydrogriseofulvin **(8)** was studied by transferring a growing culture of *Streptomyces cinereocrocatus* into a medium containing 50% deuterium oxide and undeuteriated **(8)**. Mass spectroscopy showed 0.33 deuterium atoms were incorporated per molecule of griseofulvin and deuterium NMR spectroscopy located these at C-5 and C-6 (Sato et al. 1977). Hydrogen capture in the non-oxidative conversion of squalene **(13)** to tetrahymanol **(14)** was investigated

(13)

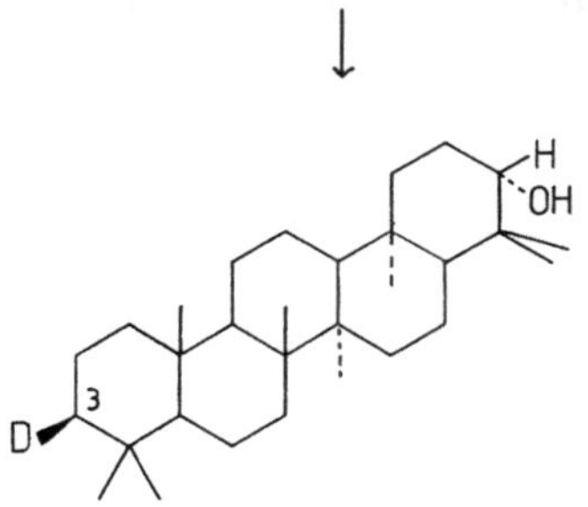

(14)

using a cell-free extract prepared from broken cells of *Tetrahymena pyriformis* (Aberhart and Caspi 1979). Because only small amounts of material are produced in cell-free transformations, it is imperative to get a compensatingly high enrichment of label in the product. This necessitated the removal of water from the system by lyophilisation and preequilibration with deuterium oxide (99.5% ^{2}H) before addition of the substrate. The tetrahymanol isolated from several incubations (11.5 mg) was monodeuteriated (41%), label residing exclusively at the 3β position.

2.2.4.3 Following Hydride Shifts

One of the most fruitful applications of ^{2}H-NMR spectroscopy to fungal systems has been in the study of hydride shifts in terpene biosynthesis. Simple single labelling experiments and more sophisticated double labelling studies have been performed. The precursors used in most of these experiments have been variously deuteriated mevalonates or acetates, with one interesting exception.

Preliminary studies with radiolabelled acetate and mevalonate showed that these precursors are not incorporated into the pentalenolactone metabolites by *Streptomyces UC5319*. However, label from [U-^{13}C]-glucose is incorporated and this prompted the use of [6,6-$^2H_2$]-glucose to act as an in vivo precursor of [2-$^2H_2$]-acetyl-CoA which, it was thought, would label H-2, H-3′, and H-4 of mevalonate. Surprisingly, the deuterium spectrum of labelled pentalenolactone E **(15)** is consistent with the incorporation of mevalonate labelled mainly at H-3′ and H-

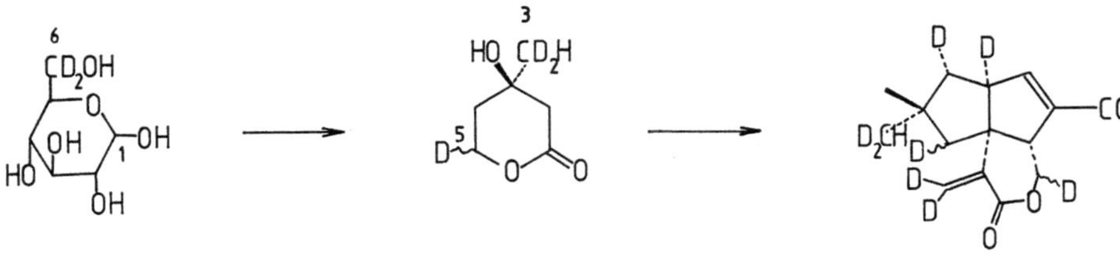

(15)

5 (Cane et al. 1981). The deuterium at C-5 of mevalonate must come from [4-^{2}H]NADH, which could have been labelled by the action of glucose-6-phosphate dehydrogenase on [1-^{2}H]-glucose. Rapid equilibration of C-1, and C-6 of hexose sugars has been demonstrated in other microorganisms (Ugurbil et al. 1978).

1,2-Hydride shifts have been suggested to account for the labelling patterns observed in studies on the biosynthesis of the structurally related sesquiterpenes PR toxin **(16)** (Gorst-Allman and Steyn 1982) and capsidiol **(17)** (Stothers 1981).

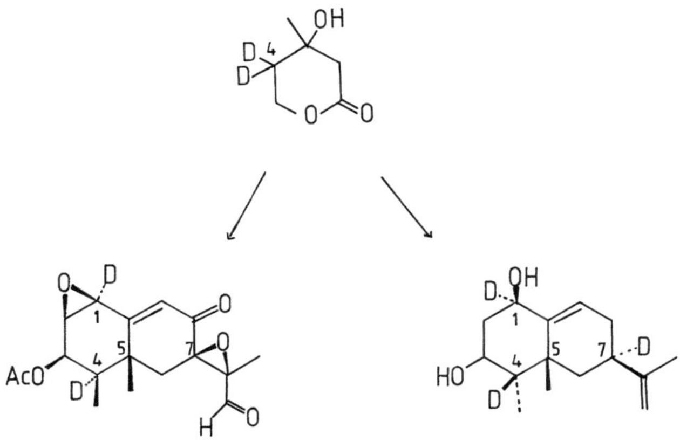

(16) (17)

[4-$^{2}H_2$]-Mevalonate labels H-1 and H-4 of PR toxin and H-1, H-4 and H-7 of capsidiol. As C-4 of mevalonate ends up at C-1, C-5 and C-7 in both cases, a hydride migration must take place.

The combined use of ^{13}C and ^{2}H labels in mevalonate enables a bond rather than an atom to be identified. Depending on whether the deuterium is directly attached to the ^{13}C nucleus in the precursor or the ^{13}C–^{2}H bond is formed in the conversion to the product, the start and terminus of a hydride migration within a mevalonate unit can be demonstrated. These strategies have been exploited in investigations of the biosynthesis of aphidicolin (Ackland et al. 1982), alliacolide (Bradshaw et al. 1982a), illudin M (Bradshaw et al. 1982c), methyl botryaloate (Bradshaw et al. 1981) and the closely related sesquiterpene dihydrobotrydial **(18)** (Bradshaw et al. 1982b).

Deuterium from C-4 of mevalonate labels H-1, H-2, and H-5 of dihydrobotrydial. C-2 of **(18)** is not derived from C-4 of mevalonate, so the deuterium on H-2 must have migrated there. Two mechanisms for this migration were envisaged, either two consecutive 1,2-hydride shifts (H-9 to C-1 and H-1 to C-2) or a direct

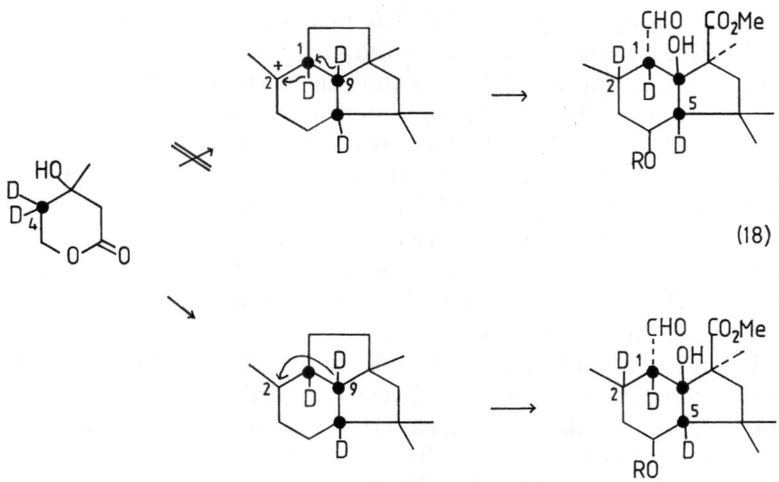

(18)

1,3-shift (H-9 to C-2). These mechanisms were distinguished when [4-^{13}C,4-^{2}H$_2$]-mevalonate was incorporated into dihydrobotrydial. Doublets (J = 20 Hz) due to heteronuclear coupling to ^{13}C in the ^{2}H-NMR spectrum showed that both H-1 and H-5 are still attached to their ^{13}C partners. The integrity of the bond to deuterium at C-1 excludes the mechanism involving two 1,2-hydride shifts which would have left only H-5 attached to its ^{13}C partner. The deuterium and ^{13}C labels at C-1 would have come from different mevalonate units and so would not both be present in the same molecule.

2.2.5 The Use of [2-^{2}H$_3$]-Acetate as a Polyketide Precursor

Extensive use has been made of ^{2}H-NMR spectroscopy to study deuterium incorporation from acetate into fungal polyketides. The reasons for the popularity of this experiment are twofold: hydrogen isotopes offer a sensitive probe to the many postulated reactions involved in assembling a polyketide; and acetate is one of the few precursors shown to be utilised by polyketide synthase enzymes. Unfortunately, detailed interpretation of many of the deuterium incorporation results is not yet possible, due to our poor understanding of the nature and mode of action of the enzyme systems involved.

[2-^{2}H$_3$]-, [1-^{13}C,2-^{2}H$_3$]- and [2-^{13}C,2-^{2}H$_3$]-acetate are all routinely used as precursors. The deuterium distribution in the metabolite can be determined by ^{2}H- or ^{13}C-NMR spectroscopy. The detection of deuterium indirectly in the ^{13}C-NMR spectrum is discussed in detail later in this review. However, it is worthwhile at this juncture to compare the two techniques. ^{2}H-NMR spectroscopy is used to show the stereochemistry of deuterium labelling and the relative distribution of deuterium at several sites in a molecule. ^{13}C-NMR spectroscopy furnishes complementary data on the multiplicity of labelling (how many deuterons are bonded to a particular carbon), and the extent of hydrogen exchange relative to ^{13}C. ^{13}C-NMR spectroscopy offers much greater resolution than ^{2}H-NMR spectroscopy.

In many investigations both techniques are used, as in a study on the biosynthesis of the pentaketide metabolite mellein **(19)** (Abell et al. 1982). Incorporation of [2-$^2H_3$]-acetate by *Aspergillus melleus* labels four positions in mellein. C-7, C-5, C-4, and C-9. The relative deuterium retention at C-4 compared to the starter methyl C-9 is 1:2.1, which suggested that a significant proportion of molecules have two deuterium atoms at the methylene position. The biosynthetic sample of mellein was methylated and the ^{2}H-NMR spectrum of the product recorded in benzene. Under these conditions, the diastereotopic hydrogens at C-4 are non-equivalent and the signals are well resolved. The two sites were equally labelled. Even together these two results are not sufficient to firmly establish that any particular molecule has two deuterium atoms at C-4. This was shown by feeding [2-^{13}C,2-$^2H_3$]-acetate, where resonances for C-4 corresponding to –CD_2– and –CDH– were observed (Abell et al. 1983).

The stereochemistry of labelling in polyketides derived from [2-$^2H_3$]-acetate has only been determined in a few cases. Insufficient chemical shift dispersion prevented a complete analysis of the labelling pattern in brefeldin **(20)** (Hutchinson et al. 1981 b) and averufin (Simpson et al. 1982). This problem was overcome by using shift reagents to show that deuterium labels only the pro-S positions of alternate carbons in the alkyl portions of 2-n-hexyl-5-n-propylresorcinol (see Sect. 2.4). In the biosynthesis of O-methylasparvenone **(21)**, deuterium atoms

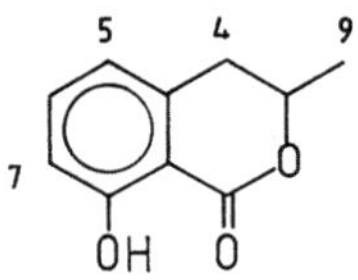

(19)

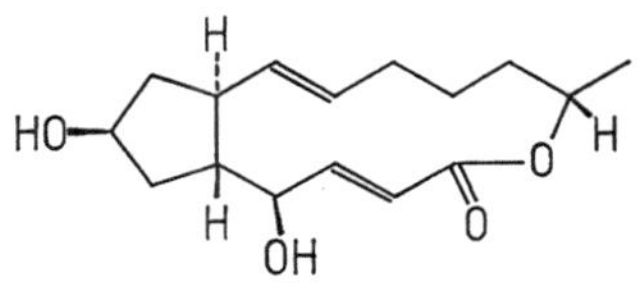

(20)

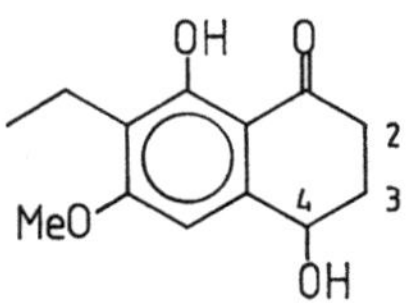

(21)

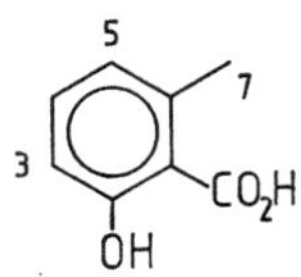

(22)

from [2-$^2H_3$]-acetate are incorporated only at the axial positions at C-2 and C-3. A precursor in which this ring is aromatised seems probable (Simpson and Stenzel 1981). The transposition of deuterium onto C-3, presumably from C-4, may occur as an NIH shift when oxygen is introduced at C-4. A similar shift is observed in the biosynthesis of aflatoxin B_1 (Simpson et al. 1983).

^{2}H-NMR spectroscopy has been used quantitatively to measure the relative deuterium distribution at different loci in a polyketide. In this way, it is often observed that more deuterium is retained per hydrogen site in the chain starter methyl group than at the malonate-derived positions. For example, in the biosynthesis of 6-methylsalicylic acid **(22)**, the relative deuterium retention from [2-$^2H_3$]-acetate at C-3, C-5, and the C-7 methyl group is 1:1.2:4.9 (Abell and Staunton 1981). Apparent exceptions to this trend were reported for the incorporation of [2-$^2H_3$]-acetate into griseofulvin **(5)** (Sato et al. 1976) and O-methylasparvenone **(21)** (Simpson and Stenzel 1981). By feeding [1-^{13}C,2-$^2H_3$]-acetate and analysing the ^{13}C-NMR spectrum, it was possible to show that the terminal methyl group of **(21)** retains at most only two deuterium atoms. This is probably also the case for the starter methyl group of griseofulvin. Although the signal for C-7 in the proton-coupled ^{2}H-NMR of enriched griseofulvin is a singlet, the resonance sharpened up and increased in height by 17% when broadband proton decoupling was applied. The failure to observe ^{1}H-^{2}H coupling is not surprising, since the theoretical line width $(\pi T_2)^{-1}$ is greater than the expected coupling of about 1.7 Hz.

A correlation has been observed between the relative deuterium retention at the two methine positions at the site of chain folding in several aromatic polyketides and the oxidation state of the carbon between those sites (Abell et al. 1982). Metabolites with a hydroxyl on carbon e (see scheme) generally retain more deuterium at position f than d, whereas for metabolites which have been reduced at position e, this order is reversed. On the basis of this observation, it was proposed that the different distributions are a result of the respective formation of either an enol or a double bond between carbons d and e orientating the three contiguous C-C bonds (marked in heavy type) in a syn arrangement. It is envisaged that keto-enol tautomerisation with non-stereospecific reprotonation on carbon depletes the amount of deuterium at position d. Implicit in this proposal is the assumption that the deuterium distribution in a series of metabolites reflects general processes, artefactual or functional, which occur on polyketide synthases.

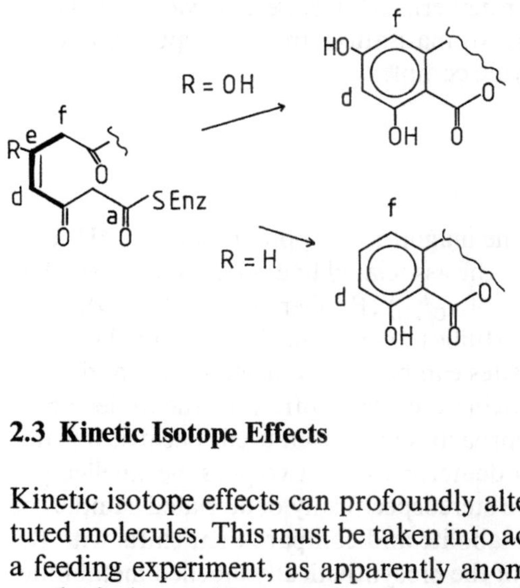

2.3 Kinetic Isotope Effects

Kinetic isotope effects can profoundly alter the reactivity of isotopically substituted molecules. This must be taken into account when interpreting the results of a feeding experiment, as apparently anomalous labelling patterns can arise. A particularly graphic example of an isotope effect was reported in a study on the biosynthesis of the phytotoxic metabolite fommanosin **(23)** (Cane and Nachbar 1980). Three atoms of tritium out of a possible six were incorporated from [2-^{14}C,5-^{3}H]-mevalonate into the sesquiterpene, however the ^{2}H-NMR of **(23)** labelled by [5-$^2H_2$]-mevalonate showed deuterium at D-10α, D-10β, and D-5 in the ratio 1:1:0.4. The discrepancy between the two sets of results, namely the less than integral retention of deuterium at D-5, arises from the use of an intramolecularly double-labelled precursor in the second experiment. When two labels are on the same molecule isotopic discrimination at one site necessarily results in

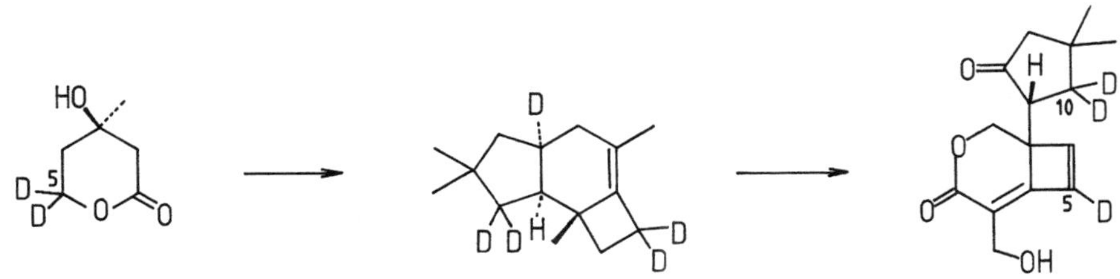

(23)

depletion of both labels. Since each molecule of an intermediate will usually contain only one labelled C_5 unit, the isotope effect causes a reduction in the relative amount of label at the affected site. In contrast, when tritiated precursors are employed no two tritium atoms reside on the same molecule, so that any isotope effect on the loss of one tritium atom from H-5 of mevalonate has no effect on the relative retention of the other tritium atom at that centre.

The fact that protium and deuterium are distinguished in biosynthetic reactions was used as the basis of a novel investigation of the aromatisation of the simple tetraketide, 6-methylsalicyclic acid **(22)** (Abell and Staunton, submitted for publication). By looking at the deuterium distribution at C-3 and C-5 of **(22)** labelled separately from mono- and trideuteriated acetates, it was possible to shaw that the reactions involved in the aromatisation are stereospecific and so presumably occur on the synthase enzyme complex.

2.4 The Use of Shift Reagents

Paramagnetic-induced shifts are the same magnitude in ppm in both the ^2H- and ^1H-NMR spectrum of a compound, but the associated line broadening is smaller for deuterium (theoretically by a factor of γ_D/γ_H) (Reuben and Fiat 1969). Shift reagents have been used in biosynthetic studies to distinguish deuterium labelling at prochiral sites in a molecule. These sites can be made diastereotopic and magnetically inequivalent either by introducing a chiral centre into the molecule or by using a chiral shift reagent. Both approaches have been used. Because it is not possible to predict the relative shifts for deuterons at the two possible labelled positions, it has been necessary to prepare stereospecifically deuteriated samples of simple degradation products of the metabolite and compare their shifts with the same product isolated by degradation of the biosynthetically labelled metabolite.

The incorporation of the reversed dimethylallyl group into the hemiterpenoid echinulin **(24)** was studied by ^2H-NMR spectroscopy using a chiral shift reagent (Harrison and Quinn 1983 a). [Me-$^2H_3$]-Mevalonate was fed to a surface culture of *Aspergillus amstelodami* and the isolated echinulin degraded to give 2,2-dimethylbutanol. It had previously been shown that the [$^2H_3$]-methyl groups of racemic [2-$^2H_3$]-methyl-2-methylbutan-1-ol can be distinguished by the differential downfield shifts induced by the chiral shift reagent Eu(hfbc)$_3$ (Harrison and Quinn 1983 b). In the presence of one equivalent of shift reagent, the methyl resonances are sufficiently resolved ($\Delta\delta = 0.066$ ppm). Analysis of the biosynthetic

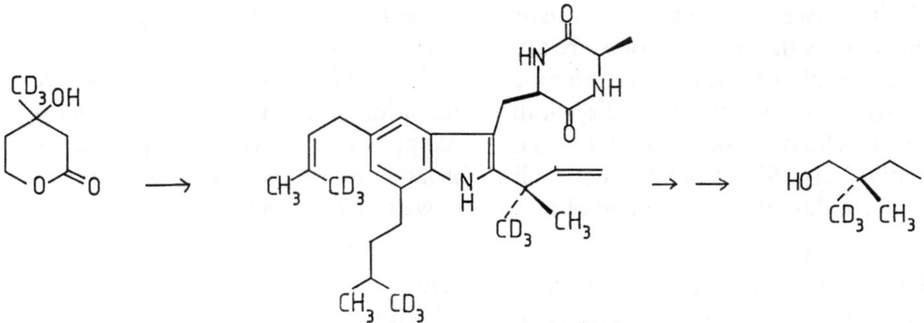

(24)

sample showed that 89% of the deuterium was in the 2-pro-S-methyl group, showing that this is the one predominantly derived from the methyl group of mevalonate.

In the biosynthesis of fatty acids and some other polyketides, alternate carbon atoms are stereospecifically labelled with deuterium from [2-$^{2}H_3$]-acetate. This labelling pattern has been analysed for 2-n-hexyl-5-n-propylresorcinol **(25)** (Furukawa et al. 1983 a). Labelled **(25)** was degraded to give heptanol and butanol. The

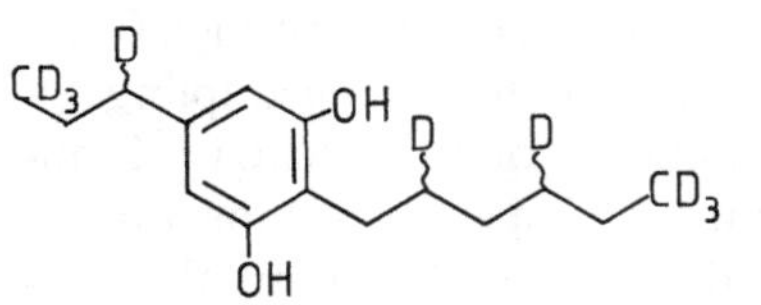

(25)

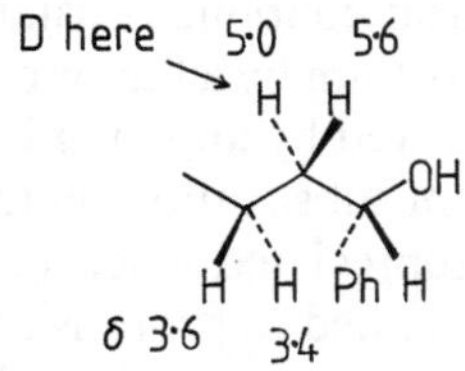

alcohols were converted to the corresponding alkylphenylcarbinols which were resolved via their 1-(−)-menthoxy carbonyl esters. Because of the newly generated asymmetry at C-1, the methylene positions on the alkyl chain are diastereotopic. This non-equivalence was enhanced in the ^{2}H-NMR spectrum by the addition of up to equimolar amounts of the lanthanide shift reagent Eu(fod)$_3$ which cleanly separated the resonances for the diastereotopic hydrogens on C-2 and C-3 of the alkylphenylcarbinols. The assignments shown were made by comparison with spectra of synthetically deuteriated samples (Furukawa et al. 1983 b).

3 Detection of Deuterium Indirectly in the ^{13}C-NMR Spectrum

3.1 Background and Techniques

The detection of deuterium by its effect on the chemical shift and multiplicity of a ^{13}C resonance in the ^{13}C-NMR spectrum enables the labelling pattern of individual molecules to be deduced. Intramolecular double-labelling with deuterium and ^{13}C is used to label bonds, measure the extent of hydrogen exchange and look at how many deuterium atoms are attached to a particular carbon. This information is complementary to that obtained from ^{2}H-NMR spectroscopy. ^{13}C-NMR spectroscopy is a much higher resolution technique than ^{2}H-NMR spectroscopy, since the chemical shift dispersion is over 200 ppm and individual resonances are usually sharp. Significantly, there is little broadening of resonances with increasing molecular size in the range of molecular weight 200–1000.

3.1.1 Detecting Deuterium Directly Attached to a Carbon Reporter Nucleus (the α-Shift Technique)

The presence of a deuterium atom directly bonded to ^{13}C is detected by the associated effects on the shielding and relaxation behaviour of the ^{13}C nucleus. A carbon bonded to one deuterium atom resonates 0.3–0.6 ppm upfield of the corresponding protonated species. Each additional isotopic replacement results in a further 0.3–0.6 ppm upfield shift. In the proton-noise-decoupled (PND) ^{13}C spectrum, heteronuclear coupling to deuterium (I = 1) increases the multiplicity of the carbon signal. –CD– appears as a triplet of equal intensity signals, –CD_2– as a quintet (intensity 1:2:3:2:1) and –CD_3 as a heptet (intensity 1:3:6:7:6:3:1). The one-bond heteronuclear coupling constant $J(^{13}C–^{2}H)$ is smaller than the corresponding coupling to protium by a factor of 6.51, and is typically around 20 Hz. This coupling can be removed by applying broadband deuterium decoupling.

The effect of deuterium substitution on the relaxation of an attached ^{13}C nucleus is often not fully appreciated when recording ^{13}C spectra of biosynthetic samples. Carbon atoms bonded to protons depend chiefly on dipole-dipole interactions with those protons for spin-lattice relaxation. In addition, the signals for protonated carbons are significantly enhanced (up to three times) when broadband ^{1}H decoupling is applied – the nuclear Overhauser effect. The magnitude of the NOE for a protonated carbon is independent of the number of directly attached protons so the intensities of the signals for –CH_2– and –CHD– provide a direct assay of the deuterium content. This comparison cannot be extended to carbons with no directly attached protons, e.g. –CD_2– or –CD–, which generally display no NOE. Spin-lattice relaxation (T_1) of carbon by deuterium is much less efficient than by protium since the rate of relaxation is proportional to the square of the gyromagnetic ratio. Consequently fully deuteriated carbons have much longer T_1 relaxation times. Deuterium, however, contributes to the spin-spin relaxation of ^{13}C, which can cause a broadening of the $^{13}C–^{2}H$ resonance.

The sensitivity of the α-shift technique is seriously reduced when observing deuteriated carbons with no directly bonded protons by the factors outlined above: poor relaxation, loss of NOE and signal multiplicity of the $^{13}C–^{2}H$ resonances in the PND spectrum. Indeed in several studies on polyketide biosynthesis, the incorporation of deuterium from [2-^{13}C,2-$^{2}H_3$]-acetate into methine positions was inferred initially only from the reduction in intensity of the corresponding $^{13}C–^{1}H$ signal. As it is frequently desirable to measure the extent of deuteriation at a particular carbon or the number of deuterons attached to a carbon, steps must be taken to improve the signal-to-noise of the $^{13}C–^{2}H$ resonances. The three most widely used approaches are:

1. application of deuterium decoupling,
2. the use of long relaxation delays and smaller pulse angles to avoid partial saturation,
3. addition of paramagnetic reagents.

These are all described in detail later.

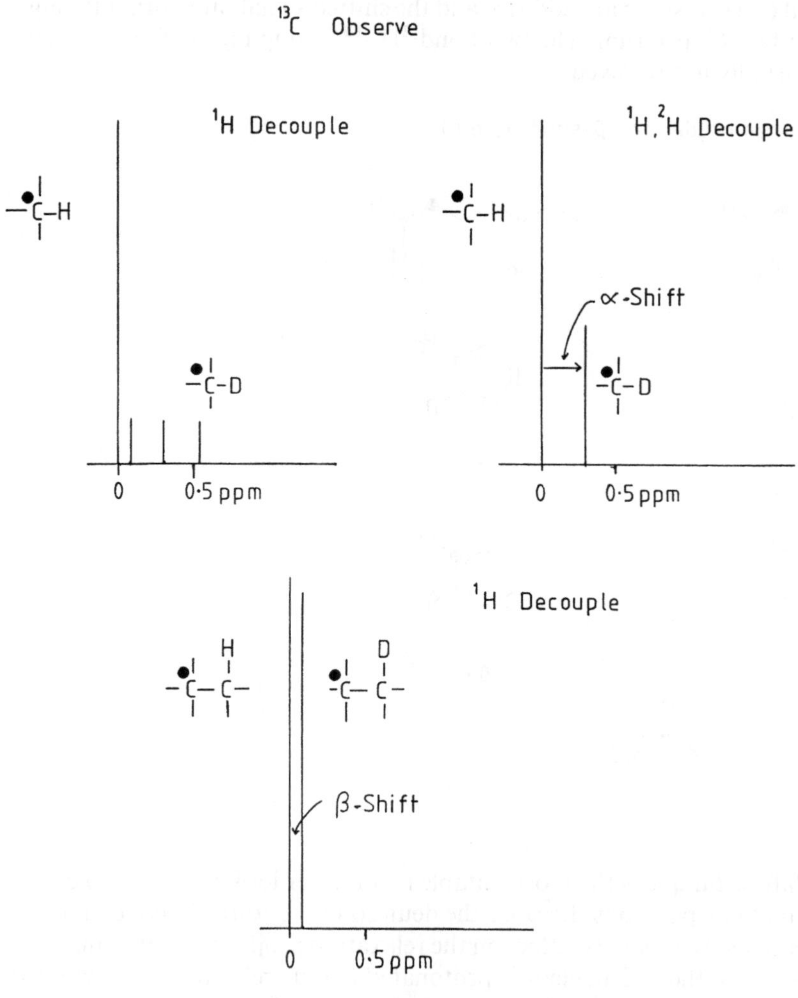

3.1.2 Detecting Deuterium on a Carbon Adjacent to the Reporter Nucleus (the β-Shift Technique)

The β-shift technique exploits the small shifts caused by isotopic replacement of protium with deuterium on a carbon adjacent to the reporter ^{13}C nucleus. β-Shifts are less pronounced and less consistent than α-shifts, being variable in both magnitude and sign. The shifts are less than 0.15 ppm and are upfield except for carbonyl carbons, which have downfield shifted resonances (Marciel et al. 1967). Studies of β-shifts in a range of aromatic systems revealed that the magnitude of the isotopic shift is a function of both the change in the normal vibrational modes of the molecule and the electronic demand of the other substituents (Bell et al. 1972). For example, a carbon bearing a methoxy substituent exhibits a much smaller β-shift (0.04 ppm) than a protonated carbon (0.1 ppm) when the adjacent aromatic position is deuteriated. Interestingly, different shifts have been reported for axial and equatorial deuterons in reduced ring systems (Simpson and Stenzel 1982 b). In all cases, β-shifts are additive and the shifted signals are normally singlets in the PND ^{13}C spectrum. The two-bond ^{2}H–^{13}C coupling is often less than 1 Hz and is usually not resolved.

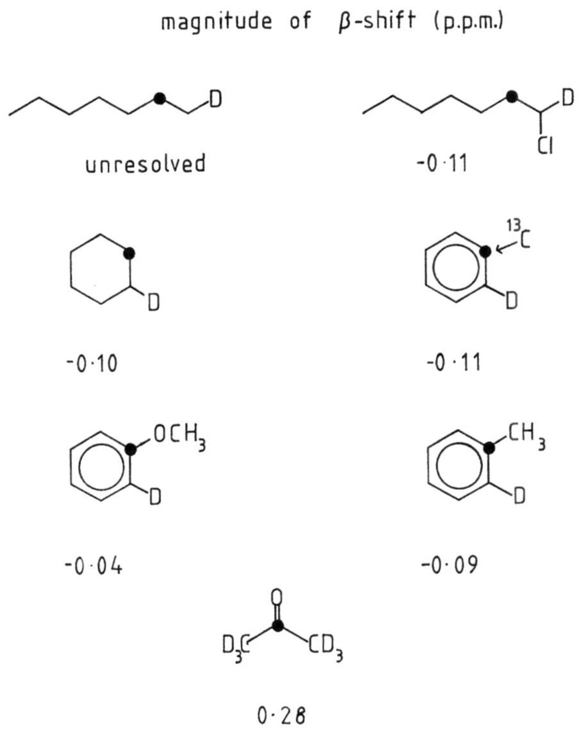

The β-shift technique is the most suitable method for looking at isotopic substitution at methine positions. Because the deuteron is not directly bonded to the ^{13}C reporter nucleus, it has less effect on the relaxation properties of that nucleus. In many instances, the ^{13}C nucleus is protonated and its relaxation behaviour is effectively independent of isotopic substitution on the neighbouring centre.

In comparing the α- and β-shift techniques, important similarities and differences are apparent. Both techniques provide information on the integrity of bonds between carbon and deuterium and the multiplicity of labelling. α-Shifts are large and of predictable magnitude, whereas β-shifts may be seen more easily in the absence of deuterium decoupling. Since deuterium decoupling is often not required in investigations using the β-shift, spectra may be acquired on standard ^{13}C probes. This technical point makes the β-shift technique attractive, especially if a triple resonance probe is not available. For quantitative experiments, the α-shift technique is superior when the reporter nucleus also retains a proton, e.g.

the –CDH– group in fatty acids and reduced polyketides. The more remote location of the reporter nucleus in the β-shift approach is, however, well suited for measuring the extent of deuteriation at fully deuteriated centres, e.g. methine positions in aromatic polyketides. The main drawback of both techniques is the low sensitivity of ^{13}C as a label. Despite the greater sensitivity of detection of ^{13}C relative to deuterium, the higher natural abundance means that smaller dilutions of precursor can be tolerated in a biosynthetic experiment.

3.2 Applications to the Study of Biosynthesis in Fungi

In the introduction to this section, specific applications of ^{13}C,^{2}H double-labelling experiments were alluded to, e.g. labelling a bond. These uses will be discussed in turn, considering the distinguishing features of each type of investigation. Many studies of deuterium distribution have, however, been less directed in their aims, seeking only to establish which deuterons derive from simple precursors, and then explaining any interesting features of the observed distribution. Counted amongst such studies are many of the experiments where new developments in technique were introduced or where a complex biosynthetic pathway is being studied and any additional data serves to consolidate the accumulated body of knowledge. These studies will be reviewed first.

Early applications of ^{13}C-NMR spectroscopy to monitor deuterium retention relied almost exclusively on the detection of deuterons directly bonded to ^{13}C. Studies into the biosynthesis of the polyketides terrein **(26)** (Garson et al. 1977 a) and scytalone (Sankawa et al. 1977) displayed the potential of incorporation experiments with [2-^{13}C,2-$^{2}H_3$]-acetate. In each case deuterium incorporation was inferred from the reduction in intensity of the ^{13}C–^{1}H resonances. By recording the spectrum of terrein without proton decoupling, the ^{13}C–^{1}H signals for C-8 were moved well away from the less intense ^{13}C–^{2}H triplet. In follow-up studies, deuterium decoupling was used to improve the signal-to-noise of the ^{13}C–^{2}H signal, and especially to detect the signal for the –CD_3 unit which identifies the start of the polyketide chain. In the deuterium-decoupled spectrum of enriched terrein, the signal for the trideuteriated starter methyl is 0.8 ppm upfield of the position for the protonated carbon and is flanked by a doublet ($J = 123$ Hz) (shifted 0.55 ppm upfield) due to molecules labelled with –CHD_2 (Garson et al. 1977 b). A similar pattern was observed for the starter unit of rugulosin **(27)** (Sankawa et al. 1978) where additionally the absence of deuterium at C-6 suggests that decarboxylation at this position follows aromatisation.

The intricate details of the complex biosynthetic pathway to the potent mycotoxin aflatoxin B_1 **(28)** are slowly being unravelled. [2-^{13}C]- and [2-^{13}C,2-$^{2}H_3$]-acetate were separately administered to a mutant strain of *Aspergillus parasiticus* and incorporated into averufin **(29)** (Sankawa et al. 1982). Comparison of the intensities of the labelled protonated carbons in the PND spectra from the two feedings enabled an approximate measure of deuterium retention at C-5, C-7, C-2′, C-4′, and C-6′ to be made. All these positions were about 40% deuteriated. The spectrum of the tribenzoate of labelled averufin showed that only one deuteron was present at each of C-2′ and C-4′. This is consistent with the side chain being

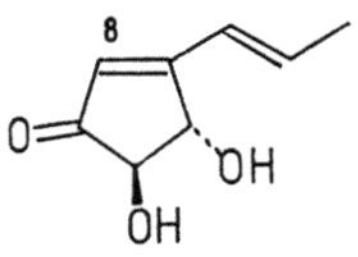

(26)

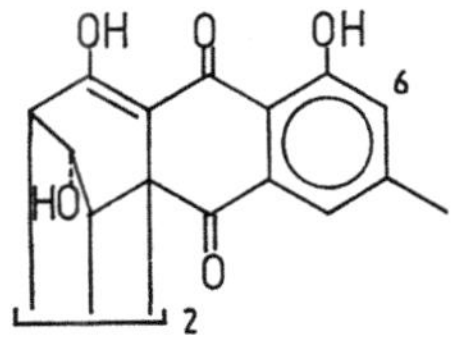

(27)

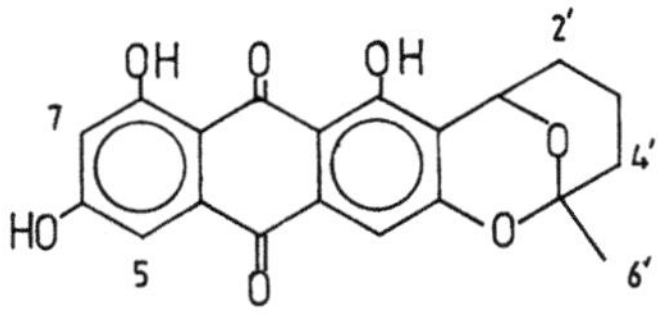

(29)

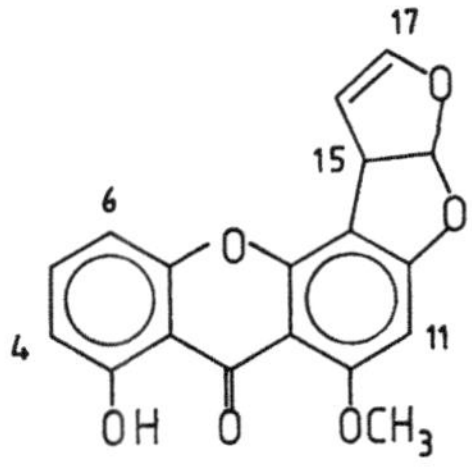

(30)

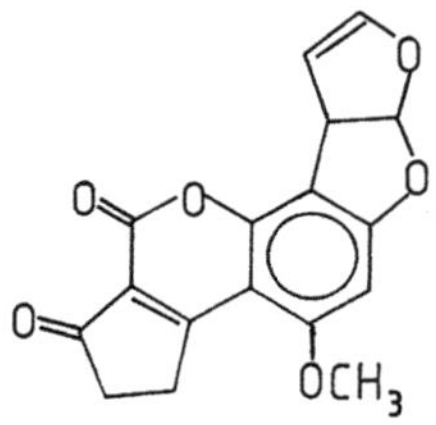

(28)

derived from a fatty acid. Deuterium incorporation into sterigmatocystin **(30)** by *Aspergillus versicolor* has been studied using both [2-^{13}C,2-^{2}H$_3$]-acetate (Sankawa et al. 1982) and [1-^{13}C,2-^{2}H$_3$]-acetate precursors (Simpson and Stenzel 1982 a). The former investigation showed deuterium retained at all the expected locations, C-4, C-6, C-11, C-15, and C-17. Surprisingly, β-shifted signals are only seen for carbons 5, 14, and 16. This may reflect a different result for the two experiments. Alternatively, the β-shifts for C-3 and C-12 may be very small, so the shifted signals are not resolved.

3.2.1 Quantitative Measurements of Deuterium Retention

Quantitative measurements of the extent of deuteriation can be made by analysis of the ^{13}C-NMR spectrum of the enriched compound if the spectrum is recorded under suitable conditions (Stothers 1974). The aim is to record a spectrum where the intensity of the signals due to $-^{13}$C–(C)–^{1}H and $-^{13}$C–(C)–^{2}H corresponds to the abundance of those species. Because the intensity of a signal depends upon the relaxation behaviour of the ^{13}C nucleus, the protonated and deuteriated carbons must be equally relaxed. Furthermore ^{13}C spectra recorded with broadband proton-noise decoupling will have different NOE enhancements for protonated and fully deuteriated carbons.

The complications caused by different rates of relaxation and different NOE's are overcome in two ways. Long relaxation delays (5–7 T_1s when using a 90 ° pulse) between pulses allow the nuclei to relax completely. Additionally, if the proton decoupler is switched off during the delay, the NOE which builds up during the acquisition time has time to decay (Freeman et al. 1972). In practice, smaller pulses (20–50 °) and correspondingly shorter delays are used. Paramagnetic relaxation reagents provide an alternative relaxation pathway for carbon and in doing so, suppress the NOE (Levy et al. 1976). Cr(acac)$_3$ is frequently used, at concentrations between 0.01 and 0.1 M. By using high concentrations of relaxation reagent, T_1s of 60 s may be reduced to 1 s. This allows rapid pulsing without saturation but may result in some signal broadening. Relaxation reagents and gated-decoupling are often combined in quantitative experiments.

Heteronuclear coupling increases the multiplicity of a signal and so reduces the signal-to-noise of each component. Because it is important to have good signal-to-noise, ideally greater than 25:1, spectra of compounds with deuterium directly attached to carbon are recorded with simultaneous proton and deuterium decoupling. Only the proton decoupling needs to be gated. Under these conditions, the signals due to both protonated and deuteriated species are singlets. When the carbon-deuterium coupling is not resolved, as is often the case for the signal due to $-^{13}C–C–^{2}H$, only gated proton decoupling is required.

The application of ^{13}C-NMR spectroscopy to measure deuterium retention at different sites in a molecule has been applied to polyketide metabolites. These molecules often retain deuterium at many similar (aromatic or alkyl) positions which cannot easily be resolved in the deuterium spectrum because of the poor spectral dispersion. This was the case in a study of the biosynthesis of palmitic acid **(31)** in *Anacystis nidulans* (McInnes et al. 1979). [2-^{13}C,2-^{2}H$_3$]-Acetate was incorporated into **(31)** and the ^{13}C spectrum of the methyl ester recorded with si-

(31)

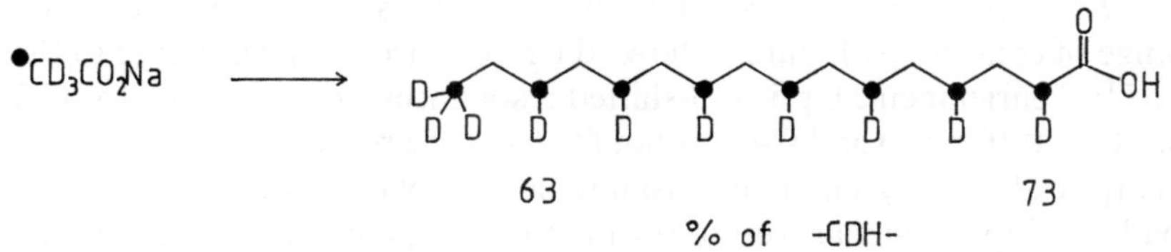

multaneous proton and deuterium decoupling. The proton decoupler was turned off for 6.8 s between pulses, Cr(acac)$_3$ (10 mg ml^{-1} of solvent) was added to reduce T_1s and a pulse angle of 51 ° was used. Under these conditions the deuterium retention at several positions in methyl palmitate was determined to a high degree of accuracy ($\pm 2\%$), and was found to be greater for carbons nearer the carboxyl terminus. A mechanistic interpretation of this result is difficult in the light of a subsequent study of fatty acids from yeast and algae where the same regular increase in deuterium retention is not observed (McInnes et al. 1983). Incorporation experiments with [2-^{13}C,2-^{2}H$_3$]-acetate showed that between 40 and 60% of a deuterium atom was retained at malonate-derived carbons in brefeldin **(20)** (Hutchinson et al. 1981a) and 2-n-hexyl-5-n-propylresorcinol **(25)** (Sankawa et

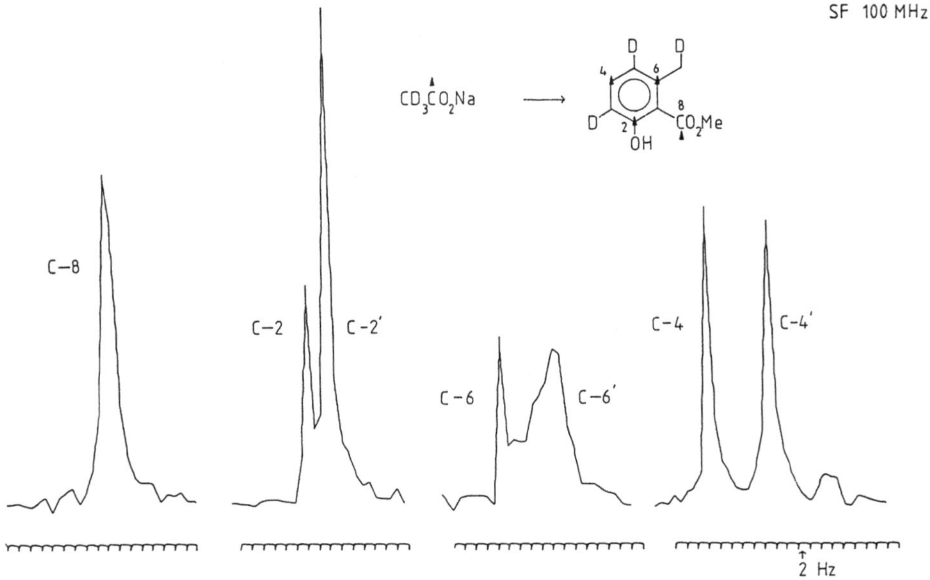

Fig. 2. Detail from the ^{13}C-NMR spectrum of methyl 6-methyl-salicylate labelled by [1-^{13}C,2-$^{2}H_3$]-acetate

al. 1981) whereas none could be detected in cytochalasins B and D (Wyss et al. 1980) and chaetoglobosin (Probst and Tamm 1981).

The limitations of quantitatively measuring deuterium retention at a methine position with the reporter ^{13}C at that position prompted an investigation into the utility of having the reporter nucleus at the neighbouring position (Abell and Staunton 1981). [1-^{13}C,2-$^{2}H_3$]-Acetate was incorporated into 6-methylsalicylic acid **(22)** by *Penicillium griseofulvum* and the ^{13}C-NMR spectrum was recorded under a range of conditions. Figure 2 shows the resonances corresponding to the four sites of ^{13}C enrichment. Upfield β-shifted resonances were observed for C-2 (0.04 ppm), C-4 (0.10 ppm) and C-6 but not for C-8. The resonances for C-6 consist of a sharp unshifted signal arising from molecules with no deuterium at C-7 and a broad envelope of signals (up to 0.1 ppm upfield) corresponding to molecules with one or more deuterium atoms at C-7. The deuterium retention at C-5 (80%) can be measured by comparison of the peak intensity of the shifted and unshifted resonances in the spectrum recorded without a long relaxation delay. This is possible because C-4 is protonated and its relaxation behaviour and NOE are effectively independent of deuteriation at C-5. In contrast, C-2 relaxes by dipole-dipole relaxation with the proton on C-3 and so is sensitive to isotopic replacement at this position. To allow for this, spectra were recorded using long relaxation delays (30 s) and small pulse angles. These conditions had been found to give an accurate measurement of the extent of deuteriation in synthetically deuteriated samples of **(22)**. In calculating the deuterium retention relative to ^{13}C it is necessary to take into account the contribution to the unshifted signal of ^{13}C at natural abundance (1.1%).

3.2.2 Determining the Multiplicity of Labelling

One of the most powerful aspects of using ^{13}C-NMR spectroscopy to look at deuterium retention is the ability to look at the labelling pattern of individual labelled species. In particular the number of deuterons attached to a carbon can be determined directly from the size of the isotopic shift and the multiplicity of the shifted signal for that carbon. In most experiments the retention of the maximum number of deuterons at a site in the labelled metabolite is the most significant result as this can preclude more oxidized intermediates. This was elegantly demonstrated in a study on the biosynthesis of the pentaketide, asperlactone **(32)** (Garson and Staunton 1981). [2-^{13}C,2-$^{2}H_3$]-Acetate was incorporated by *Aspergillus melleus* into **(32)**, and the ^{13}C spectrum recorded with simultaneous proton and deuterium decoupling. The spectrum features four signals for the methyl group at C-10 and three for C-7. The signals due to C-10 correspond to molecules of asperlactone with $-CH_3$, $-CDH_2$, $-CD_2H$, and $-CD_3$ in the methyl group respectively, whereas only the first three labelling patterns are observed for C-7. This information identifies C-10 as part of the starter acetate unit and furthermore precludes any aromatic precursors such as **(33)** which could retain at most only one deuteron at C-7 of asperlactone.

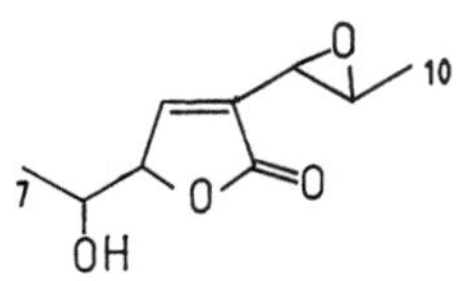

(32)

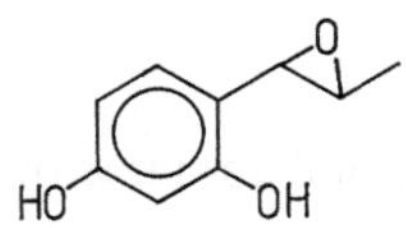

(33)

A problem which has recently been the focus of much attention is the establishment of the precursors in macrolide and polyether biosynthesis. One specific question is whether S- or R-methylmalonyl CoA, or both epimers, are used to furnish the three carbon units used by the synthase to extend the polyketone chain. By labelling the two epimers at C-2 with deuterium and monitoring the incorporation into product it should in principle be possible to answer this question. [2-^{13}C,2-$^{2}H_2$]-Propionate serves as a precursor of 2S-[2-^{13}C,2-^{2}H]methylmalonyl CoA **(34)** and [2-^{13}C,2-$^{2}H_2$]succinate labels the 2R compound **(35)** at C-2 with deuterium. Although **(34)** and **(35)** are readily interconvertible enzymically the epimerisation proceeds with complete loss of hydrogen from C-2. Thus deuterium from C-2 of propionate can only label centres on the polyketone chain with a D configuration, and succinate those with an L configuration. So far deuterium from labelled propionate has been detected at the predicted centres in two systems, lasalocid A (Hutchinson et al. 1981c) and erythromycins A and B **(36)** (D.E. Cane, personal communication 1984). In the latter study, deuterium was detected only in the methylene of the propionate starter unit in the spectrum recorded with simultaneous proton and deuterium decoupling. However, in the difference spectrum obtained by subtracting spectra recorded with the deuterium decoupler switched on and off respectively, additional isotopically shifted signals (0.38–0.45 ppm upfield) were observed for C-2, C-4, and C-10, whereas no deuterium was present at C-8 and C-12. The use of such difference spectra enhances

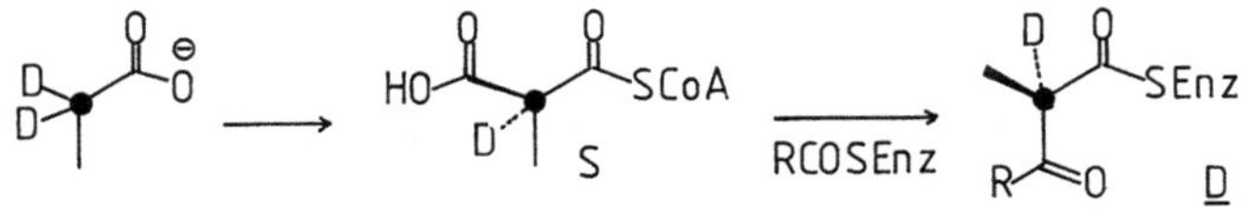

(34)

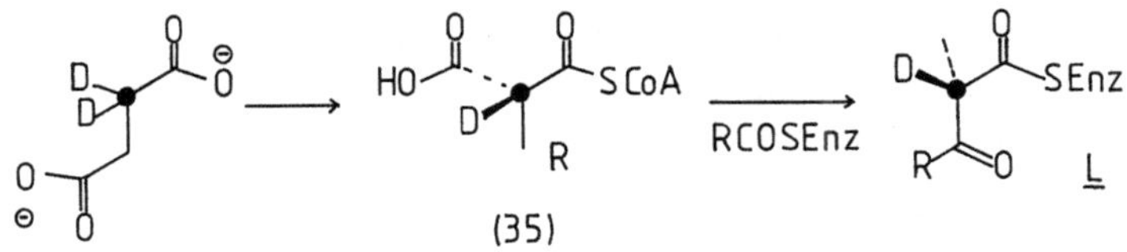

(35)

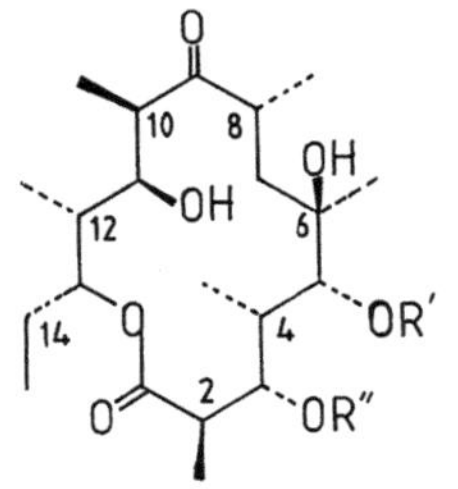

(36)

the sensitivity of detection of small amounts of deuterium (a few percent relative to ^{13}C).

3.2.3 Labelling a Bond

The perturbation to a ^{13}C resonance by isotopic substitution with deuterium on or adjacent to the reporter nucleus effectively labels the intervening bonds, and can be used to monitor bond making or breakage in biosynthesis. Several examples of the use of doubly labelled precursors to follow hydride shifts in terpene biosynthesis are described in Section 2.2.4.3. In these examples ^{2}H-NMR is used to detect one-bond heteronuclear coupling. ^{13}C-NMR spectroscopy is potentially more useful, being able to relate isotopic labels not directly attached to each other and so label more than one bond. This is illustrated in studies of the biosynthesis of ascochlorin (Hunter and Mellows 1978) and fusicoccin (Banerji et al. 1978).

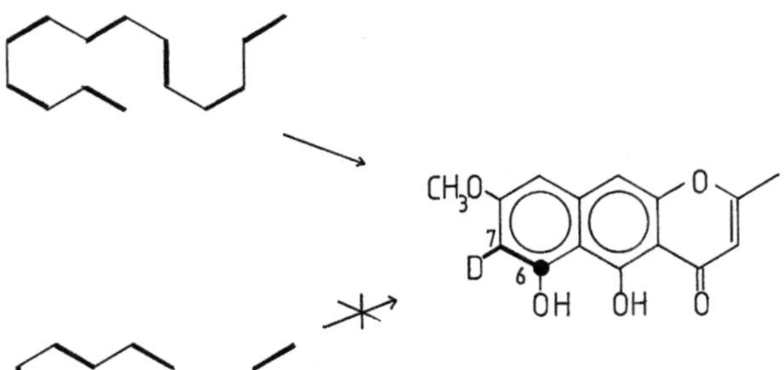

(37)

$[1,2\text{-}^{13}C_2]$-Acetate is an invaluable aid in determining the folding pattern of the polyketide chain in many metabolites. Similarly, the detection of a β-shifted resonance in a metabolite enriched with $[1\text{-}^{13}C,2\text{-}^{2}H_3]$-acetate shows an intact carbon-carbon bond (as part of a $^{13}C\text{–}C\text{–}^{2}H$ unit) and so can be used to determine the orientation of the C_2 unit. The folding pattern of the polyketone chain of the heptaketide, rubrofusarin **(37)**, was confirmed using this precursor (Leeper and Staunton 1982). The β-shifted signal at C-6 in the PND ^{13}C spectrum of enriched **(37)** must be due to molecules labelled with deuterium at C-7 and therefore C-6 and C-7 must come from the same acetate unit.

4 Conclusions

This review has illustrated the range of experiments that have been performed to investigate fungal metabolism by detecting deuterium using NMR spectroscopy. It should be appreciated that the decade from 1974–1984 has seen this study reach a level of sophistication and maturity which has enabled a rapid advance in our knowledge of biosynthetic processes. Not only has NMR spectroscopy proved the ideal technique to investigate fungal metabolism, but inevitably the requirements of the biosynthetic investigations have extended the scope of the spectroscopic techniques. Indeed, combined with the use of NMR spectroscopy to monitor oxygen incorporation, via isotopic shifts in the ^{13}C-NMR spectrum (Vederas 1982), and C-N bond integrity from the heteronuclear coupling (Gould 1982), the bioorganic chemist now has the means to establish the origin of most atoms in a fungal product.

In the years ahead the continued development of the techniques described in this review will depend on their applicability to studies, with cell-free systems or purified enzymes, of individual transformations on a biosynthetic pathway. One consequences of this move away from studying biosynthesis in the intact fungus may be the increased use of tritium NMR (Elvidge 1982) instead of deuterium NMR for direct observation of a hydrogen isotope. The excellent nuclear magnetic properties of tritium have not been fully exploited so far because of the need to use high levels of radioactivity in precursors. However, as more powerful spectrometers become available, so the minimum detectable level of activity will drop to very manageable levels.

Acknowledgements. I would like to thank Dr. K. Abell, Dr. D. E. Cane, Dr. E. D. Laue and Dr. F. J. Leeper for their respective contributions to the preparation of this review.

References

Abell C, Staunton J (1981) The use of ^{2}H-NMR spectroscopy and β-isotopic shifts in the ^{13}C-NMR spectrum to measure deuterium retention in the biosynthesis of the polyketide 6-methylsalicyclic acid. J Chem Soc Chem Commun 856–858

Abell C, Garson MJ, Leeper FJ, Staunton J (1982) Biosynthesis of the fungal metabolites alternariol, mellein, rubrofusarin, and 6-methylsalicylic acid from CD_3CO_2H. J Chem Soc Chem Commun 1011–1013

Abell C, Doddrell DM, Garson MH, Laue ED, Staunton J (1983) Biosynthesis of the polyketide mellein from CD_3CO_2H and $^{13}CD_3CO_2H$ in *Aspergillus melleus*: detection of deuterium by ^{2}H-NMR and edited ^{13}C-NMR spectra. J Chem Soc Chem Commun 694–696

Aberhart JD, Caspi E (1979) Nonoxidative cyclization of squalene by *Tetrahymena pyriformis*. Incorporation of a 3β hydrogen (deuterium) atom into tetrahymanol. J Am Chem Soc 79:1013–1019

Ackland MJ, Hanson JR, Ratcliffe AH, Sadler IH (1982) A ^{2}H-NMR study of the rearrangement step in aphidicolin biosynthesis. J Chem Soc Chem Commun 165–166

Baldwin JE, Jung M, Usher JJ, Abraham EP, Huddleston JA, White RL (1981) Penicillin biosynthesis: conversion of deuteriated (L-α-amino-δ-adipyl)-L-cysteinyl-D-valine into isopenicillin N by a cell-free extract of *Cephalosporium acremonium*. J Chem Soc Chem Commun 246–247

Banerji A, Hunter R, Mellows G, Sim K-Y, Barton DHR (1978) Use of deuterium as a tracer with ^{13}C nuclear magnetic resonance spectroscopy in following deuteride migration in terpenoid biosynthesis: mechanism of geranylgeranyl pyrophosphate cyclisation in fusicoccin biosynthesis. J Chem Soc Chem Commun 843–845

Barber J, Staunton J (1980 a) Biosynthesis of citrinin in *Penicillium citrinum*: advanced precursor studies using ^{2}H nuclear magnetic resonance spectroscopy. J Chem Soc Chem Commun 552–554

Barber J, Staunton J (1980 b) Citrinin biosynthesis in *P. citrinum*: incorporation of a deuterium-labelled advanced precursor in the presence of a specific inhibitor of an early step. J Chem Soc Chem Commun 1163–1164

Bardshiri E, Simpson TJ (1983) ^{13}C and ^{2}H labelling studies on the biosynthesis of scytalone in *Phialaphora lagerbergii*. Tetrahedron 3539–3542

Battersby AR, Edington C, Fookes CJR, Hook JH (1982) Biosynthesis of vitamin B_{12}: preparation of specifically deuteriated heptamethyl dicyanocobyrinate for study by ^{2}H-NMR spectroscopy. J Chem Soc Chem Commun 181–182

Bell RA, Chan CL, Sayer BG (1972) Deuterium isotope effects on ^{13}C chemical shifts in benzene and substituted benzenes. J Chem Soc Chem Commun 67–68

Bradshaw APW, Hanson JR, Sadler IH (1981) Use of ^{2}H–^{13}C-NMR coupling patterns in terpenoid biosynthesis. J Chem Soc Chem Commun 649–650

Bradshaw APW, Hanson JR, Sadler IH (1982 a) Hydrogen rearrangements in the biosynthesis of the sesquiterpenoid, alliacolide. J Chem Soc Chem Commun 292–293

Bradshaw APW, Hanson JR, Nyfeler R, Sadler IH (1982 b) Studies in terpenoid biosynthesis. Part 25. The fate of the mevalonoid hydrogen atoms in the biosynthesis of the sequiterpenoid, dihydrobotrydial. J Chem Soc Perkin Trans I:2187–2192

Bradshaw APW, Hanson JR, Sadler IH (1982 c) Studies in terpenoid biosynthesis. Part 26. Applications of ^{2}H and ^{13}C-NMR spectroscopy to the biosynthesis of the illudin sesquiterpenoids. J Chem Soc Perkin Trans I:2445–2448

Bradshaw APW, Hanson JR, Sadler IH (1982 d) Studies in terpenoid biosynthesis. Part 27. The acetate and mevalonate labelling patterns of the sesquiterpenoid, alliacolide. J Chem Soc Perkin Trans I:2787–2790

Bycroft BW, Wels CM, Corbett K, Lowe DA (1975) Incorporation of [α-^{2}H]- and [α-^{3}H—-L-cystine into penicillin G and the location of the label using isotope exchange and ^{2}H nuclear magnetic resonance. J Chem Soc Chem Commun 123

Cane DE, Murthy PPN (1977) Application of deuterium magnetic resonance to biosynthetic studies. 2. Rosenonolactone biosynthesis and stereochemistry of a biological S_N2' reaction. J Am Chem Soc 99:8327–8329

Cane DE, Nachbar RB (1980) Biosynthesis of fommanosin. ^{2}H-NMR evidence against postulated hydride shifts. Tetrahedron Lett 437–440

Cane DE, Rossi T, Tillman AM, Pachlatko JP (1981) Stereochemical studies of isoprenoid biosynthesis. Biosynthesis of pentalenolactone from [U-$^{13}C_6$]glucose and [6-$^2H_2$]glucose. J Am Chem Soc 103:1838–1843

Damtoft S (1981) Biosynthesis of lamiide and ipolamiide from 8-*epi*-deoxyloganin studied by ^{2}H-NMR spectroscopy. J Chem Soc Chem Commun 228–229
Dewick PM, Ward D (1977) Stereochemistry of isoflavone reduction during pterocarpin biosynthesis: an investigation using deuterium magnetic resonance spectroscopy. J Chem Soc Chem Commun 338–339
Diehl P, Leipert Th (1964) Deuteronen-Kernresonanzspektroskopie. Helv Chim Acta 47:545–557
Egan W (1976) Intermolecular association in adenosine 5′-monophosphate. An ^{2}H nuclear magnetic resonance investigation. J Am Chem Soc 98:4091–4093
Elvidge JA (1982) Tritium NMR spectroscopy and its applications. In: Duncan WP, Susan AB (eds) Synthesis and applications of isotopically labeled compounds. Elsevier, Amsterdam, pp 35–44
Erlenmeyer H, Schoenauer W, Sullmann H (1936) Chemische und biochemische Dehydrierung einer Anthan-α-d,α′-d-dicarbonsäure. Helv Chim Acta 19:1376–1380
Freeman R, Hill HDW, Kaptein R (1972) Proton-decoupled NMR spectra of carbon-13 with nuclear overhauser effect suppressed. J Magn Res 7:327–329
Furukawa J, Iwasaki S, Okuda S (1983 a) Eu(fod)$_3$-shifted ^{2}H-NMR as a probe of chirality due to deuterium substitution: stereospecific deuterium incorporation into 2-n-hexyl-5-n-propylresorcinol, a polyketide produced by *Pseudomonus* sp. B-9004. Tetrahedron Lett 5261–5264
Furukawa J, Iwasaki S, Okuda S (1983 b) Method for determination of enantiomeric composition and absolute configuration of 2,3-deuterated 3-alkylpropanols. Tetrahedron Lett 5257–5260
Garson MJ, Staunton J (1979) Some new NMR methods for tracing the fate of hydrogen in biosynthesis. Chem Soc Rev 8:539–560
Garson MJ, Staunton J (1981) The biosynthesis of asperlactone: incorporation studies with [2-^{13}C,2-^{2}H$_3$)acetate. J Chem Soc Chem Commun 708–710
Garson MJ, Hill RA, Staunton J (1977 a) Deuterium as a tracer in polyketide biosynthesis: incorporation of [2-^{13}C,2-^{2}H$_3$]acetate into terrein. J Chem Soc Chem Commun 624–626
Garon MJ, Hill RA, Staunton J (1977 b) Deuterium as a tracer in polyketide biosynthesis: a new method for the detection of chain starter units. J Chem Soc Chem Commun 921–922
Glasel JA, Hruby VJ, McKelvy JF, Spatola AF (1973) Deuteron magnetic resonance studies on the microdynamical behaviour of partially deuterated oxytocin with neurophysin. J Mol Biol 79:555–557
Gorst-Allman CP, Steyn PS (1982) Biosynthesis of PR toxin by *Penicillium roqueforti*, part 2 evidence for an hydride shift form ^{2}H-NMR spectroscopy. Tetrahedron Lett 5359–5362
Gould SJ (1982) Biosynthesis of natural products – an overview of current results and problems. II. Studies of nitrogen-containing metabolites. J Nat Prod 45:38–49
Harrison DM, Quinn P (1983 a) The biosynthesis of echinulin: origins of the diastereotopic methyls in the 1,1-dimethylallyl group. J Chem Soc Chem Commun 879–880
Harrison DM, Quinn P (1983 b) The stereospecific synthesis of (S)-2-[^{2}H$_3$]methyl-2-methylbutanol. Characterisation of the (R) and (S) enantiomers of the racemic [^{2}H$_3$]-alcohol by ^{2}H-NMR in the presence of a chiral shift reagent. Tetrahedron Lett 831–834
Hunter R, Mellows G (1978) Detection of deuteride shifts in the biosynthesis of the fungal triprenylphenol, ascochlorin, by ^{13}C nuclear magnetic resonance spectroscopy following incorporation of [3-^{13}C,4-^{2}H$_2$]mevalonic acid. Tetrahedron Lett 5051–5054
Hutchinson CR (1982) The use of isotopic hydrogen and nuclear magnetic resonance spectroscopic techniques for the analysis of biosynthetic pathways. J Nat Prod 45:27–37
Hutchinson CR, Heckendorf AH, Straughn JL, Daddona PE, Cane DE (1979) Biosynthesis of camptothecin. 3. Definition of strictosamide as the penultimate precursor assisted by ^{13}C and ^{2}H-NMR spectroscopy. J Am Chem Soc 101:3358–3369
Hutchinson CR, Kurobane I, Mabuni CT, Kumola RW, McInnes AG, Walter JA (1981 a) Biosynthesis of macrolide antibiotics. 3. Regiochemistry of isotopic hydrogen labeling of brefeldin A by acetate. J Am Chem Soc 103:2474–2477

Hutchinson CR, Kurobane I, Cane DE, Hasler H, McInnes AG (1981 b) Biosynthesis of macrolide antibiotics. 4. Stereochemistry of hydrogen labeling of brefeldin A by [2-$^{2}H_3$]acetate. J Am Chem Soc 103:2477–2480
Hutchinson CR, Sherman MM, McInnes AG, Walter JA, Vederas JC (1981 c) Biosynthesis of macrolides. 6. Mechanism of stereocontrol during the formation of lasalocid A. J Am Chem Soc 103:5956–5959
Leeper FJ, Staunton J (1982) Rubrofusarin biosynthesis in *Fusarium culmorum*: incorporation of $^{13}CH_3^{13}CO_2H$ and $CD_3^{13}CO_2H$ into the polyketide napthalene nucleus. J Chem Soc Chem Commun 911–912
Levy GC, Edlund U, Holloway CF (1976) Mechanisms for interactions between organic molecules and paramagnetic relaxation reagents. J Magn Res 24:375–387
Marciel GE, Ellis PD, Hofer DC (1967) Carbon-13 chemical shifts of the carbonyl group. V. Observation of a deuterium isotope effect using carbon-13 field-frequency lock. J Phys Chem 71:2160–2164
McInnes AG, Walter JA, Wright JLC (1979) Differential hydrogen exchange during the biosynthesis of fatty acids in *Anacystis nidulans*: the incorporation of [2-$^{2}H_3$,2-^{13}C]acetate. Tetrahedron Lett 3245–3248
McInnes AG, Walter JA, Wright JLC (1983) Regiochemical distribution of deuterium during fatty acid biosynthesis following incorporation of [2-^{13}C,2-$^{2}H_3$]acetate. Tetrahedron 39:3515–3522
Probst A, Tamm C (1981) 199. Biosynthesis of the cytochalasans. Biosynthetic studies on chaetoglobosin A and 19-*O*-acetylchaetoglobsin A. Helv Chim Acta 64:2065–2077
Rana J, Robins DJ (1983) Pyrrolizidine alkaloid biosynthesis; incorporation of ^{2}H-labelled putrescines into retroorsine. J Chem Soc Chem Commun 1222–1224
Reuben J, Fiat D (1969) On the deuteron magnetic resonance in solutions of paramagnetic ions. J Am Chem Soc 91:1242–1243
Sankawa U, Shimada H, Sato T, Kinoshita J, Yamasaki K (1977) Biosynthesis of scytalone. Tetrahedron Lett 483–486
Sankawa U, Shimada H, Yamasaki K (1978) Acetate-[2-^{13}C,2-$^{2}H_3$] as a precursor in polyketide biosynthesis: detection of deuterium incorporation with ^{13}C-NMR. Tetrahedron Lett 3375–3378
Sankawa U, Shimada H, Yamasaki K (1981) Biosynthesis of 2-hexyl-5-propylresorcinol: biosynthetic incorporation of deuterium from [2-^{13}C,2-$^{2}H_3$]acetate. Chem Pharm Bull 29:3601–3605
Sankawa U, Shimada H, Kobayashi T, Ebizuka Y, Yamamoto Y, Noguchi H, Seto H (1982) Biosynthesis of averufin and sterigmatocystin in *Aspergillus parasiticus* and *A. versicolor*: application of multiple labelled acetate in polyketide biosynthesis. Heterocycles 19:1053–1058
Sato Y, Oda T, Saito H (1976) A novel biosynthetic study of griseofulvin by ^{2}H nuclear magnetic resonance: determination of deuterium incorporation from [2-$^{2}H_3$]acetate by *Penicillium urticae*. Tetrahedron Lett 2695–2698
Sato Y, Oda T, Saito H (1977) Microbial transformation of dehydrogriseofulvin and griseofulvin: ^{2}H-NMR and mass spectrometric studies of stereochemical courses of microbial hydrogenation and hydroxylation. J Chem Soc Chem Commun 415–417
Sato Y, Oda T, Saito H (1978) ^{2}H nuclear magnetic resonance on biosynthesis: stereochemistry of the 5′-hydrogen atoms of griseofulvin derived from griseophenone B and 4-demethyldehydrogriseofulvin. J Chem Soc Chem Commun 135–136
Shaha AJ, Keeler J, Freeman R (1983) Evaluation of a new broadband decoupling sequence: waltz 16. J Magn Res 53:313–340
Simpson TJ, Stenzel DJ (1981) ^{13}C and ^{2}H-NMR studies on the biosynthesis of O-methylasparvenone, a hexaketide metabolite of *Aspergillus parvulus*. J Chem Soc Chem Commun 239–240
Simpson TJ, Stenzel DJ (1982 a) Biosynthesis of aflatoxins. Incorporation of [1,2-$^{13}C_2$]acetate, [$^{2}H_3$]acetate, and [1-^{13}C,$^{2}H_3$]acetate into sterigmatocystin in *Aspergillus versicolor*. J Chem Soc Chem Commun 890–892
Simpson TJ, Stenzel DJ (1982 b) Application of ^{2}H β-isotopic shift in ^{13}C-NMR spectra to biosynthetic studies. Incorporation of [1-^{13}C,$^{2}H_3$]acetate into O-methylasparvenone in *Aspergillus parvulus*. J Chem Soc Chem Commun 1074–1076

Simpson TJ, de Jesus AE, Steyn PS, Vleggaar R (1982) Biosynthesis of aflatoxins. Incorporation of [2-$^{2}H_3$]acetate and [1-^{13}C,2-$^{2}H_3$]acetate into averufin. J Chem Soc Chem Commun 632–634
Simpson TJ, de Jesus AE, Steyn PS, Vleggaar R (1983) Biosynthesis of aflatoxins. Incorporation of [2-$^{2}H_3$]acetate into aflatoxin B_1 by *Aspergillus flavus*. J Chem Soc Chem Commun 338–340
Stothers JB (1974) ^{13}C-NMR studies of reaction mechanisms and reactive intermediates. In: Levy GC (ed) Topics in carbon-13 NMR spectroscopy, vol. 1, Wiley-Interscience, New York, pp 231–244
Stothers JB (1981) Sesquiterpenes – biosynthetic studies with ^{13}C and ^{2}H magnetic resonance – a synthetic approach via monoenolization. Pure Appl Chem 53:1241–1258
Townsend CA, Brown AM, Nguyen LT (1983) Norcardicin A: stereochemical and biomimetic studies of monocyclic β-lactam formation. J Am Chem Soc 105:919–927
Turner WB (1971) Fungal metabolites. Academic Press, London
Turner WB, Aldridge DG (1983) Fungal metabolites II. Academic Press, London
Ugurbil K, Brown TR, den Hollander JA, Glynn P, Shulman RG (1978) High-resolution ^{13}C nuclear magnetic resonance studies of glucose metabolism in *Escherichia coli*. Proc Natl Acad Acid Sci USA 75:3742–3746
Vederas JC (1982) Biosynthetic studies using ^{18}O shifts in ^{13}C nuclear magnetic resonance. Can J Chem 60:1637–1642
Wyss R, Tamm C, Vederas JC (1980) 161. Differential hydrogen exchange during biosynthesis of cytochalasins B and D. Helv Chim Acta 63:1538–1541

Proton NMR Studies on DNA Structure

P. Bendel

1 Introduction

Although the most detailed and accurate determinations of structural features in nucleic acids rely upon X-ray studies in crystals and fibers, Nuclear Magnetic Resonance has increasingly gained in importance during the last decade, as a tool for conformational studies of DNA. One of the basic reasons for this success is the fact that NMR succeeds where X-ray studies fail, namely in aqueous solutions (the more natural environment for DNA) and in the localization of hydrogen atoms.

The rapid progress in this field during the last years was sparked off by several factors. One was the availability of high-field NMR instruments with greatly increased resolving power, enabling routine studies at proton resonance frequencies of up to 500 MHz. This technical progress was complemented by the development of sophisticated methods for the assignment of resonances in complex molecules, most notably two-dimensional NMR (Bax 1982). The other contributing factor was the development of efficient methods to synthesize short nucleic acid oligomers of virtually any sequence, or to isolate biologically significant fragments by genetic engineering methods.

The methodology used for structure determination by proton NMR spectroscopy basically consists of: (a) the analysis of chemical shifts and coupling constants, and (b) the determination of relative (and in some cases absolute) interproton distances by use of the NOE (Nuclear Overhauser Effect). The background for both methods will be presented, along with several selected examples to highlight both the methods and the important conclusions gained from proton NMR.

The chapter is limited to the topics included in its title, namely *Proton NMR Studies of DNA Structure*. Not included are studies of other nucleic acids, such as, for example transfer-RNA, which has nevertheless produced some of the most elegant demonstrations of the power of proton NMR for structure elucidation in solution (Robillard 1979; Johnston and Redfield 1979). NMR spectroscopy of other nuclei, such as ^{31}P, has also proven fruitful in characterizing various structural features in DNA (Cohen et al. 1983). Finally, it should be kept in mind that the definition of a rigid "structure" is not necessarily sufficient to characterize DNA molecules in solution. Many of the parameters measured by NMR could very well be time-averaged values representing rapid structural fluctuations, i.e., molecular dynamics, which could be as important to the functional specificity of the DNA molecule as the structure itself (James et al. 1983).

2 Theoretical Background

2.1 Chemical Shifts

There are several main features, through which the double helix formation is manifested in the chemical shifts of the proton NMR spectrum (Kearns 1977). One is the effect of hydrogen bonding, which causes a large downfield shift of the imino protons involved in the Watson-Crick hydrogen bonds, namely the protons on N(3) of thymine and N(1) of guanine, which are shifted to a spectral range between 12 and 14 ppm downfield from TMS (Hilbers 1979). The fact that only hydrogen-bonded imino protons resonate in this spectral range, and that each base pair is represented by a single line makes these resonances extremely useful for assignments, as well as for structural and dynamic studies. The mere fact that these proton resonances can be observed in solution indicates that they are trapped by base pairing. Dynamic measurements actually show that there is an exchange of position between the imino protons and the solvent protons, but this exchange is slow on the NMR time scale. Nevertheless, imino-proton resonances can only be observed in H_2O, not in D_2O solutions. Upon raising the temperature, the exchange rate increases and the imino-proton resonances broaden, until they disappear at the onset of complete melting of the double helix.

The other important influence on the shielding constants for protons in nucleic acids is due to the ring current effects of the aromatic rings of the bases. Since in the ordered DNA structure the bases are stacked more or less parallel to each other, most proton resonances will experience upfield shifts due to the ring current effects of the bases from above and below. These shifts will disappear upon melting and conversion to a random-coil conformation.

A structure determination by chemical shifts is, in principle, based upon the measurement of various chemical shifts for resonances which can be assigned to specific protons, and comparison of this data to a set of calculated values. Each known geometry assumable by the double helix is characterized by a set of distances and angles, usually obtained from X-ray crystallographic studies, and this geometry can, in turn, be used to calculate the predicted shifts. The drawback of this approach is that it is, in principle, only as good as the accuracy of the chemical shift calculation. The basis for most calculations are the ring current contour maps of Giessner-Prettre and Pullman (1970), usually taking into account the effects from neighbor and nearest-neighbor bases (Arter and Schmidt 1976). Refinements of the ring current shift calculations take into account the diamagnetic component of the atomic magnetic anisotropy of the bases, and the anisotropic diamagnetic and paramagnetic components of the sugar-phosphate backbone (Mitra et al. 1981).

2.2 Coupling Constants

The magnitude of vicinal coupling constants can be empirically related to the dihedral angle between the protons in a H-C-C-H fragment (Karplus 1969). In nucleic acids and their fragments this information can be related to conformational

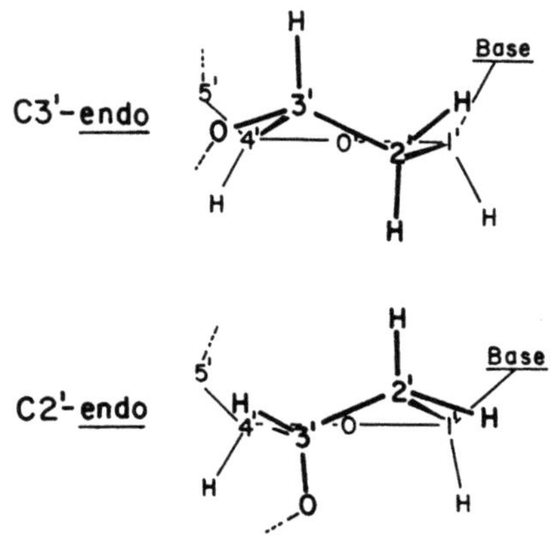

Fig. 1. Schematic representation of the two deoxyribose sugar puckering conformations encountered in A-DNA and B-DNA. Note the large distance between the base and the H3′ proton in the C2′-endo conformation. (Dickerson et al. 1982)

features in the molecule, in particular the sugar-ring conformation (Dhingra and Sarma 1979), which is characteristically different for different helical geometries. Thus, the preferred sugar conformation is C2′-endo in B-DNA, but C3′-endo in A-DNA (Fig. 1), (see Dickerson et al. 1982 for an excellent summary of A-, B-, and Z-DNA geometries). In practice, however, two or more sugar conformations can exist in dynamic equilibrium, giving rise to averaged values of coupling constants. Methods have been proposed to determine the relative weight of different conformations in the equilibrium (Evans and Sarma 1974; Lee and Tinocco 1980; Cheng et al. 1982), and used to determine conformations in small model sequences (Mellema et al. 1984).

The coupling constants among the various sugar protons are not always easy to determine directly, and usually spectral simulations by computer are necessary to obtain the relevant coupling constants. An interesting method to deduce these values from the intensity modulations obtained in a spin-echo experiment was demonstrated by Feigon et al. (1983).

Finally, it should be mentioned that heteronuclear coupling constants, such as between ^{1}H and ^{13}C, or ^{1}H and ^{31}P, are also indicative of conformational states of the sugar and the sugar-phosphate backbone (Davies 1983).

2.3 NOE

The Nuclear Overhauser Effect is observed in systems of spins with mutual dipolar intractions, provided this dipolar mechanism makes a dominant contribution to their spin-lattice relaxation (Noggle and Schirmer 1971; Bothner-By 1979) (this will usually limit it to nuclei with spin 1/2, since for $I > 1/2$ the quadrupolar mechanism is usually dominant). If, for example, nuclei i and j are dipolar coupled, they will share common energy levels and a perturbation of the magnetization of spin i will influence the transition intensity of j, and vice versa. This is the Nuclear Overhauser Effect (NOE).

NOE is usually observed by selectively irradiating the resonance frequency of, say, spins j (thereby equalizing the populations in the upper and lower energy levels of the j-transition) and recording the intensity of resonance i, compared to its intensity in the absence of the prior irradiation of j. The ratio of intensities for these two cases (i.e., with and without pre-irradiation) is called the Nuclear Overhauser Enhancement, denoted η_i. If i and j are both protons, it is given by:

$$\eta_i = \frac{\sigma_{ij}}{\varrho_j}(1 - e^{-\varrho_j t}). \tag{1}$$

Here ϱ_j is the selective spin-lattice relaxation rate of spin i, i.e., the longitudinal relaxation rate ($1/T_1$) observed for spins if if they are selectively saturated or inverted (while spins i remain at thermal equilibrium). σ_{ij} is the so-called cross-relaxation rate, and t the length of the pre-irradiation time of spins j. For sufficiently long pre-irradiation times ($t >> 1/\varrho_j$), one obtains the equilibrium NOE value of:

$$\eta_{ij} = \frac{\sigma_{ij}}{\varrho_j}. \tag{2}$$

σ_{ij} and ϱ_j are functions of the spectral densities of molecular motions, and of the strength of the mutual dipolar interaction, which in turn is proportional to the reciprocal sixth power of the internuclear distance ($1/r_{ij}^6$). In the limit of fast molecular motion, such that:

$$\omega_o^2 \tau_c^2 \ll 1, \tag{3}$$

where ω_o is the larmor frequency and τ_c the correlation time of molecular motion, the relaxation rates obey the equation:

$$\varrho_j = 2\sigma_{ij}, \tag{4}$$

so that $\eta_{ij} = 0.5$, i.e., there is a 50% enhancement of the intensity of i upon irradiation of j. Moreover, the NOE will be independent of the internuclear distance r_{ij}.

In large molecules, molecular motions are slow, and the expression in Eq. (3) is much larger than 1. In this case, the equilibrium enhancement will approach a limit of -1, i.e., the absorption of i will be wiped out upon sufficiently long irradiation of j. The cross-relaxation rate in the slow-motional limit can be expressed as:

$$\sigma_{ij} = \frac{\gamma^4 \hbar^2}{10 r_{ij}^6} \tau_c, \tag{5}$$

where γ is the gyromagnetic ratio, and $\hbar$ Planck's constant. It is therefore possible to determine η_{ij} and ϱ_j by independent experiments and calculate the cross-relaxation rate σ_{ij}. Then the internuclear distance can be determined if τ_c can be estimated. In any case in the same molecule, the build-up rate of the NOE (i.e., ϱ_j) will be inversely proportional to r_{ij}^6, so that qualitative conclusions about relative internuclear distances can by reached by measuring the NOE for short pre-irradiation times. In practice, NOE's will be difficult or impossible to detect for distances larger than about 3.5 Å, so that the detection of a NOE between two protons unequivocally demonstrates a proximity of less than this distance.

The capability to determine distances between protons is obviously of utmost importance in structure determination. Some structures can be identified (or ruled out) by the magnitude of a single distance between two protons, for which examples will be given later. In these cases a single NOE experiment will be sufficient, provided both protons are identified in the spectrum.

The assignment process itself is, of course, greatly aided by NOE measurements. In this case, the presence or absence of a NOE may have to be established among a large number of resolved lines in the spectrum, a procedure which would be very time-consuming if carried out by individual selective irradiation of each resonance, sequentially. Here, the two-dimensional version of the NOE experiment (Jeener et al. 1979; Kumar et al. 1980), called NOESY, is extremely useful, since it displays, in a single 2-D spectrum, all the NOE connectivities at once.

2.4 Assignments

It is clear that the common prerequisite for all methods of structure determination is a spectral assignment of the involved protons that is as complete and accurate as possible. We will not elaborate on the details of the various assignment procedures, but only outline the principles. In the first place, assignments for the monomeric and dimeric constituents are available through the well-established methods of chemical shifts, coupling patterns, selective deuterations or deuterium exchange, and other methods (Ts'o et al. 1969; Schweizer et al. 1968; Chan and Nelso 1969; Cheng and Sarma 1977). The effect of base pairing and base stacking has been carefully monitored and catalogued for a variety of short oligonucleotide sequences (Cheng et al. 1982; Doornbos et al. 1983; Tran-Dinh et al. 1982). As the DNA chain length increases, more resonances from protons of the same type crowd into the same spectral regions, and the resolution of individual resonances requires high-field spectrometers. The assignment of resonances from protons of the same type to the various bases along the chain requires sequential assignment methods in which a proton on a given residue is identified, followed by identification of other protons on the same monomeric unit (by probing to which other peaks the identified proton is connected by scalar coupling), and by identifying neighboring residues (by probing for spatial proximities via NOE experiments). While such measurements can be carried out sequentially, i.e., by irradiating one proton resonance at a time and looking for effects of either decoupling or NOE at other positions in the spectrum, the efficient way of performing these assignments is of course by two-dimensional experiments, in which the entire network of connectivities, either through scalar couplings or through NOE, is apparent at once. Several systematic approaches of such assignment procedures have been described and demonstrated (Hare et al. 1983; Scheek et al. 1984; Lown et al. 1984; Kan et al. 1982; Feigon et al. 1983b; Sanderson et al. 1983; Pardi et al. 1983). The largest fragments with complete assignments of most proton resonances include 17 base pairs (Weiss et al. 1984a, b; Chou et al 1983).

3 Examples

An elegant demonstration of the power of NOE experiments to elucidate solution structures of double-stranded deoxyribonucleic polymers can be found in recent publications on poly(dA-dT). This polymer has been studied quite extensively by fiber X-ray diffraction methods, and the data from these studies were interpreted by various models (Arnott and Huckins 1972; Arnott and Selsing 1974; Klug et al. 1979), including a proposed left-handed helix with Hoogsten-type base pairing (Drew and Dickerson 1982). Assa-Munt and Kearns (1984) obtained the 2-D NOE spectrum of poly(dA-dT), and were able, after assigning the proton resonances, to determine relative distances between various proton pairs, based upon the relative intensities of the appropriate NOE cross-peaks, and the length of the mixing times necessary for their appearance. They found a strong intranucleotide interaction between the aromatic AH8 (or TH6) protons and the H2′ sugar proton, whereas interactions with the H1′ proton were extremely weak or absent. This clearly points to an anti conformation around the glycosilic bond (Fig. 2). A slightly weaker (compared to the intranucleotide aromatic to H2′ interaction) cross-relaxation was detected to the H2″ proton of the neighboring nucleotide (3′–5′). All this experimental evidence is compatible only with a right-handed B-type conformation, which was further corraborated by the fact that both the A and T sugars have a 2′-endo (3′-exo) S-conformation, as deduced from the absence of cross-peaks between H3′ and the base protons (AH8 or TH6) (Fig. 1).

The base-pairing structure of poly(dA-dT) in solution was studied by a "truncated" NOE experiment (TNOE), in which the resonance frequency of the imino proton was irradiated for different times and the magnetization transfer to the two aromatic protons of A (AH2 and AH8) monitored (Assa-Munt et al. 1984). The proton which experiences the more rapid build-up of magnetization transfer is the one which is closer to the imino proton. From the experimental data (Fig. 3) this is clearly the AH2 proton, as expected for the Watson-Crick geometry of base-pairing. By contrast, the Hoogsten geometry, which was proposed for poly(dA-dT) in fibers, predicts the opposite situation, i.e., a closer proximity between the imino proton and the AH8 proton. Thus, the base-pairing structure of poly(dA-dT) in solution at 0.1 M NaCl is most probably of Watson-Crick geometry. A similar observation was reported for poly(dI-dC) (Kearns et al. 1983).

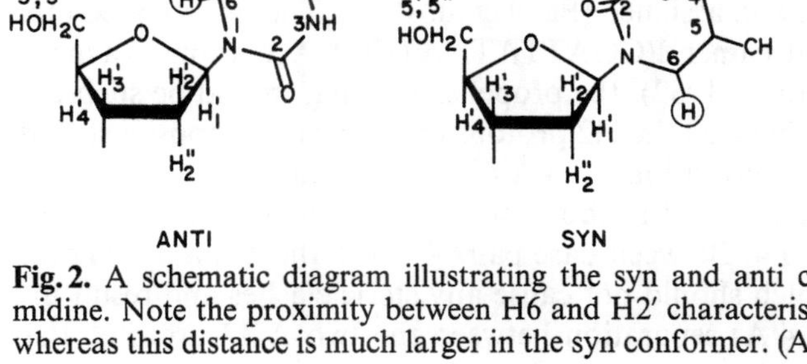

Fig. 2. A schematic diagram illustrating the syn and anti conformations for deoxythymidine. Note the proximity between H6 and H2′ characteristic of the anti conformation, whereas this distance is much larger in the syn conformer. (Assa-Munt et al. 1984)

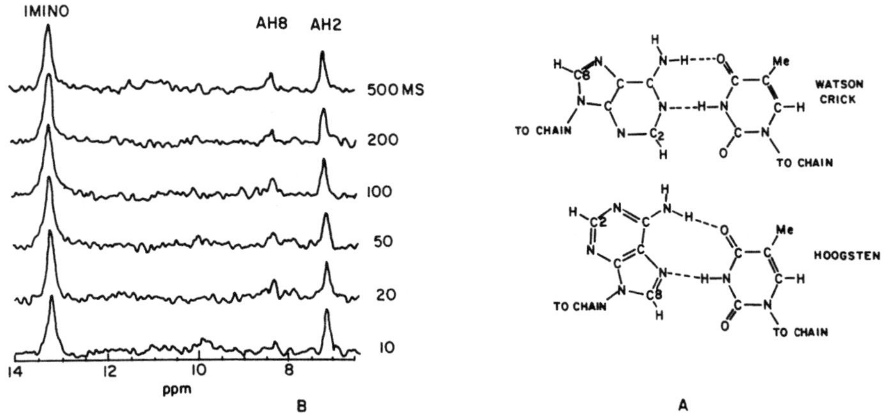

Fig. 3. A Proposed base pair structures for poly(dA-dT). Watson-Crick and Hoogsten geometry. **B** TNOE difference spectra resulting from irradiation at 13.17 ppm for the times indicated at the *right* of the figure. The transfer of magnetization to the AH2 proton (at 7.07 ppm) is clearly faster, indicating a Watson-Crick geometry. (Assa-Munt et al. 1984)

In some cases, the relative distances among different protons can be deduced from explicit measurements of their relaxation rates $1/T_1$ or $1/T_2$. These rates are expected to increase for shorter interproton distances. Feigon et al. (1983 a) measured the proton relaxation rates in the synthetic decamer d(ATATCGATAT) and found that the spin-spin relaxation rate for the resonance representing the A(3)H2 and A(9)H2 protons was faster then the rate for the other AH2 protons (on base pairs 1 and 4). This result can be rationalized by a cross-strand dipolar interaction between the AH2 protons on bases A(9) and A(3) (i.e., the second and third base pairs from the chain ends) which is stronger than for the other adenine H2 pairs on opposite strands. The interproton distance necessary to quantitatively account for the enhanced relaxation rates was calculated to be 2.8 ± 0.2 Å, which is consistent with a propelling of the bases, such that the bases on opposite strands tilt toward each other rather than staying perpendicular to the long helix axis (Dickerson and Drew 1981). A NOE measurement could not be made in this case since the two protons in question have overlapping chemical shifts, at a 300 MHz proton resonance frequency.

Another study dealing with the propeller twisting in alternating purine-pyrimidine sequences was conducted at 500 MHz, and at this high field all the AH2 resonances could be resolved and assigned (Patel et al. 1983). The investigated molecule was the synthetic dodecamer d(CGATTATAATCG). According to a mechanism proposed by Calladine (1982), the propeller twisting should be such that in a purine-(3′-5′)-pyrimidine step the H2 protons of purines on opposite strands are more distant than the conventional 3.4 Å obtained if the bases were exactly perpendicular to the helix axis. In the above molecule this would be the step between the base pairs 3 and 4. Between base pairs 4 and 5 there is a pyrimidine-(3′-5′)-pyrimidine step which should not cause any steric clashes and result in a more or less standard (3.4 Å) separation between the two AH2 protons. Base

Table 1. Steady-state NOE values (η_{ij}), selective spin-lattice relaxation rates (ϱ_j), cross-relaxation rates (σ_{ij}), and interproton distances (r_{ij}) between AH2 protons on adjacent dA–dT base pairs in d(CGATTATAATCG). (Patel et al. 1983)

Pair	Spin i	Spin j	η_{ij}	$\varrho_j (s^{-1})$	$\sigma_{ij} (s^{-1})$	r_{ij}[a] (A)
3,4	4	3	−0.09	0.97	−0.09	4.22
4,5	4	5	−0.15	1.63	−0.25	3.56
5,6	5	6	−0.38	1.31	−0.50	3.17

[a] Distances were calculated assuming $\tau_c = 9$ ns

pairs 5 and 6 represent a pyrimidine-(3′-5′)-purine step, for which a steric clash of the 2-positions on opposite-strand purines is predicted, resulting in a shorter than usual A(5)H2-A(6)H2 distance.

These predictions were all confirmed in the study of Patel et al. (1983). They also measured the cross-relaxation rates and steady-state NOE values among the various AH2 pairs, and calculated the interproton distances [Eq. (5)], assuming a correlation time for molecular motion of 9 ns (Early et al. 1980). These results are shown in Table 1. While the exact computed values of the distances depend upon the choice of τ_c, the differences between the distances is to a reasonable degree independent of this parameter, and these differences agree very well with a model derived from fiber diffraction patterns, which assumes a 13 ° propeller twist between base pairs (Arnott et al. 1983).

Another demonstration of the adaption of a propeller-twisted structure in solution by A-T base pairs, was made by Sarma et al. (1981 b), who investigated the octamer d(GGAATTCC) and dodecamer d(CGCGAATTCGCG). They based their structure determination upon chemical shifts, by computing the expected shielding constants for several published X-ray structures, and comparing them to the experimental results. Based on this comparison for five different protons on the internal A-T base pairs, they concluded that the experimental results were compatible only with a propeller-twisted structure, as predicted from theoretical considerations by Levitt (1978).

While the examples mentioned so far, and numerous others (Hare et al. 1983; Kan et al. 1982; Sanderson et al. 1983; Scheek et al. 1983; Wemmer et al. 1984; Clore and Gronenborn 1984 to name just a few) basically confirmed the features expected for a right-handed B-type geometry, other studies have adressed themselves to more unusual conformations, most notably the left-handed Z-DNA.

A study confirming the adaptation of a left-handed Z-DNA structure by poly(dG-dC) in high-salt solution was published by Patel et al. (1982). One of the characteristic features of this structure is the different glycosidic torsion angle adapted by the guanine and cytidine residues, the orientation being anti for C and syn for G. The syn orientation is characterized by a relatively short (2.2 Å) interproton distance between the aromatic proton GH8 (or CH6) and the anomeric

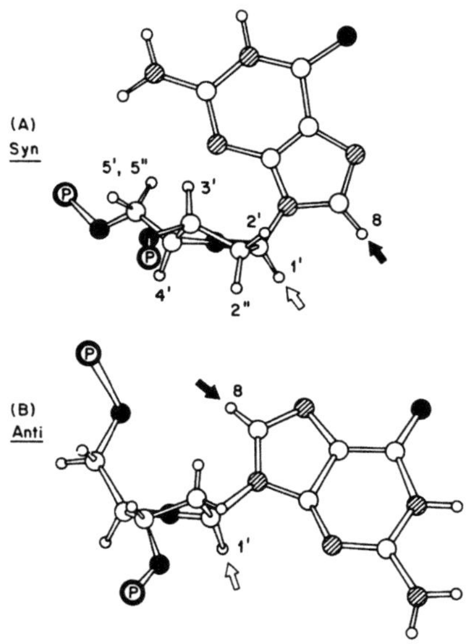

Fig. 4 A, B. Schematic representation of the syn and anti orientations about the guanosine glycosilic bond. The internuclear distance between the H8 and H1′ protons is 2.2 Å in the syn conformation, and 3.75 Å in the anti conformation. (Patel et al. 1982)

proton H1′, while this distance is much larger (3.75 Å) in the anti conformation (Fig. 4).

We thus expect to be able to differentiate between the two possibilities by measuring the NOE between the anomeric proton and the corresponding aromatic proton on the same residue, anticipating a much stronger effect for the syn conformation. The experimental results indeed demonstrated the effects expected for Z-DNA at 4 M NaCl (Fig. 5), and for B-DNA in the low-salt solution (Fig. 6). Similar observations were made for poly(dG-m^{5}dC), except that the Z-DNA conformation was completed at a lower salt concentration (1.5 M NaCl) as expected from circular dichroism measurements (Behe and Felsenfeld 1981).

The transition of poly(dG-dC) between B-DNA in low salt (without propeller twisting) and Z-DNA in high salt was also described by Sarma et al. (1981 a), based solely on chemical shift data. The measured chemical shifts of five different protons were compared to values calculated for ten different proposed DNA geometries. Only the basic B-DNA model of Arnott and Huckins (1972) yielded acceptable overall agreement with the experimental data in low-salt solution, whereas the data in high salt were entirely consistent only with the left-handed Z-DNA geometry.

The question of whether short d(G–C) sequences adopt a left-handed Z-DNA structure in high-salt solutions has been the subject of some controversy. Based upon the NOE between H8 and various sugar protons, the presence of Z-DNA was not detected for the decamer d(G-C)$_5$, which apparently retained the B-type geometry even at 4 M NaCl (Reid et al. 1983). Circular dichroism data seem to confirm this conclusion, which is in contrast to the crystal structure, where both the tetramer and the hexamer are left-handed.

There is also some ^{31}P NMR evidence pointing to a structural transition in high salt for (dC-dG)$_4$ and (dC-dG)$_3$ (Holak et al. 1984). Although the authors

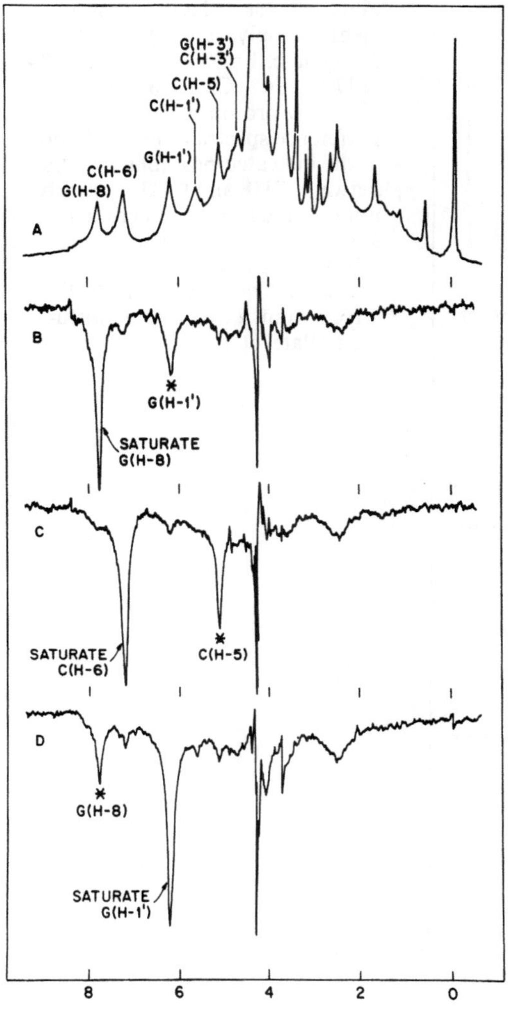

Fig. 5. Transient NOE measurements on poly(dG-dC) in 4 M NaCl, 10 mM cacodylate, 1 mM EDTA, pH 6.52, in D_2O at 45 °C. *A* Normal 498 MHz proton spectrum. *B–D* Difference spectra obtained after irradiation at the indicated positions, showing negative NOE's between GH8 and GH1′, and between CH6 and CH5. (Patel et al. 1982)

interpreted this as the formation of Z-DNA, the change in the backbone torsional angles evident from the ^{31}P spectra is strictly only indicative of the presence of a dinucleotide rather than mononucleotide repeating symmetry unit (Patel et al. 1979; Cohen et al. 1981). Such a structural feature could also be part of a so-called "alternating" B-DNA model (Klug et al. 1979).

In another series of investigations of odd structural features in DNA, Patel et al. (1982a, b, c, d) synthesized the basic Dickerson dodecamer d(CGCGAATTCGCG) (Wing et al. 1980) with different "defects", and compared the proton NMR characteristics of the perturbed structures to those of the intact molecule (Patel et al. 1982d). Replacement of an interior G-C pair by a mismatched G-T pair (Patel et al. 1982a), revealed a wobble-pair formation involving both imino protons of the mismatched pair, similar to the G-U pairs ob-

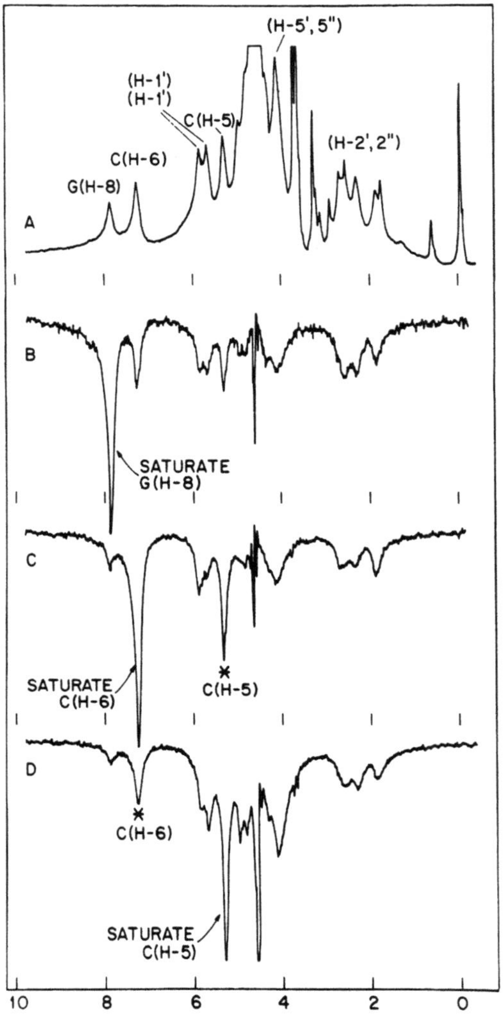

Fig. 6. Transient NOE measurements on poly(dG-dC) in 10 mM cacodylate, 1 mM EDTA, pH 7.24, in D_2O at 45 °C. *A* Normal 498 MHz proton spectrum. *B–D* Difference spectra after irradiation at the indicated positions. A NOE between GH8 and GH1′ is clearly not observed. A NOE between CH6 and CH5 is present. The appearance of other negative peaks in these spectra indicate secondary NOE effects, transferred by spin diffusion. (Patel et al. 1982)

served for several t-RNA's in solution (Johnston and Redfield 1979). The study also revealed that the conformational change associated with the incorporation of the G-T mismatch was localized to its immediate vicinity, extending one to two base pairs away from the mismatched site. Another study examined the influence of an extra adenosine residue (with no base-pairing partner on the opposite strand), between bases 3 and 4 (Patel et al. 1982b). The chemical shifts of the aromatic and imino protons revealed that the extra adenosine base was fully stacked into the duplex, and that the base-pairing for the adjacent bases was not ruptured.

4 Summary

It is, in this author's opinion, quite clear that nuclear magnetic resonance spectroscopy, in particular proton NMR, is the most powerful tool to determine details of DNA structure in solution. In the last few years, we have witnessed the development and refinement of methods and their application to the elucidation of structures for short to medium DNA fragments as a function of base sequence and solution conditions. In many of these studies the basic structural features assumed by DNA in fibers or crystals were confirmed in solution: the basic B-DNA geometry for alternating G-C sequences in low salt, Z-DNA structure for these sequences in high salt, B-DNA geometry with propeller-twisting for alternating A-T sequences, and so on. This may not be the most exciting news, but the results are nevertheless important because they established the confidence that the methods of measurement and interpretation are reliable, and can be used in studies in which problems of biological relevance are tackled more directly. These are primarily structural and dynamic studies on sequences which perform known biological functions within natural DNA, and interaction between DNA and other molecules, particularly intercalating drugs and proteins. Many studies representing this trend are already in progress, including spectral assignments in the λ operator site O_L1 (Weiss et al. 1984a, b), the phage λ O_R3 sequence (Chou et al. 1983), the Pribnow promoter sequence (Wemmer et al. 1984; Chou et al. 1984), lac-repressor lac-operator interactions (Buck et al. 1983) and EtBr intercalation (Delbarre et al. 1983). The methods of proton NMR structure determination of DNA will thus hopefully lead to many biologically significant and interesting results, concerning the function of DNA in biological processes.

References

Arnott S, Huckins DWL (1972) Optimal parameters for A-DNA and B-DNA. Biochem Biophys Res Commun 47:1504–1509

Arnott S, Selsing E (1974) Structures for the polynucleotide complexes Poly(dA). Poly(dT) and Poly(dT). Poly(dA). Poly(dT). J Mol Biol 88:509–521

Arnott S, Chandrasekaran R, Hall IH, Puigjaner LC, Walker JK, Wang M (1982) DNA secondary structures: Helices, wrinkles and junctions. Cold Spring Harbor Symp Quant Biol 47:53–65

Arter DB, Schmidt PG (1976) Ring current shielding effects in nucleic double adic helices. Nucl Acids Res 3:1437–1447

Assa-Munt N, Kearns DR (1984) Poly(dA-dT) has a right-handed B conformation in solution: a two-dimensional NMR study. Biochemistry 23:791–796

Assa-Munt N, Granot J, Behling RW, Kearns DR (1984) ^{1}H-NMR relaxation studies of the hydrogen-bonded imino protons of Poly(dA-dT). Biochemistry 23:944–955

Bax A (1982) Two-dimensional nuclear magnetic resonance in liquids. Delft University Press, Delft

Behe M, Felsenfeld G (1981) Effects of methylation on a synthetic polynucleotide: The B-Z transition in Poly(dG-M^{5}dC). Poly(dG-m^{5}dC). Proc Natl Acad Sci USA 78:1619–1623

Bothner-By AA (1979) Nuclear Overhauser effects on protons and their use in the investigation of structures of biomolecules. In: Shulman RG (ed) Biological applications of magnetic resonance. Academic, pp 177–219

Buck F, Hahn KD, Zemann W, Rueterjans H, Sadler JR, Beyreuther K, Kaptein R, Scheek R, Hull WE (1983) NMR study of the intercalation between the lac repressor and the lac operator. Eur J Biochem 132:321–327

Calladine CR (1982) Mechanism of sequence-dependent stacking of bases in B-DNA. J Mol Biol 161:343–352

Chan SI, Nelson JH (1969) Proton magnetic resonance studies of ribose dinucleoside monophosphates in aqueous solution. I. The nature of the base-stacking interaction in adenylyl (3′-5′)adenosine. J Am Chem Soc 91:168–183

Cheng DM, Sarma RH (1977) Nuclear magnetic resonance study of the impact of ribose 2′-o-methylation on the aqueous solution conformation of cytidylyl-(3′-5′)-cytidine. Biopolymers 16:1687–1711

Cheng DM, Kan LS, Leutzinger EE, Jayaraman K, Miller PS, Ts'o POP (1982) Conformational study of two short pentadeoxyribonucleotides, d-CpCpApApG and d-CpTpTpGpG and their fragments by proton nuclear magnetic resonance. Biochemistry 21:621–630

Chou SH, Hare DR, Wemmer DE, Reid BR (1983) Sequence-specific recognition of deoxyribonucleic acid. Chemical synthesis and nuclear magnetic resonance assignment of the imino protons of λ O_R3 operator deoxyribonucleic acid. Biochemistry 22:3037–3041

Chou SH, Wemmer DE, Hare Dr, Reid BR (1984) Sequence specific recognition of DNA: NMR studies of the imino protons of a synthetic RNA polymerase promoter. Biochemistry 23:2257–2262

Clore GM, Gronenborn AM (1984) A nuclear overhauser enhancement study of the solution structure of a double-stranded DNA undecamer comprising a portion of the specific target size for the cyclic AMP-receptor protein in the gal operon. Eur J Biochem 141:119–129

Cohen JS, Wooten JB, Chatterjee CL (1981) Characterization of alternating deoxyribonucleic acid conformations in solution by phosporus-31 nuclear magnetic resonance spectroscopy. Biochemistry 20:3049–3055

Cohen JS, Chen CW, Knop RH (1983) Secondary structure of DNA in solution by NMR methods. In: Pullman B, Jortner J (eds) Nucleic acids: The vectors of life. D. Reidel, pp 127–139

Davies DB (1983) Dependence of oligonucleotide conformational properties on sugar ring and base-sequence. ^1H and ^{13}C-NMR studies. In: Pullman B, Jortner J (eds) Nucleic acids: The vectors of life. D Reidel, pp 141–154

Delbarre A, Gourevitch MI, Gaugain B, Le Pecq JB, Roques BP (1983) ^1H-NMR study of an ethidium dimer Poly(dA-dT) complex: Evidence of a transition between bis and monointercalation. Nucl Acids Res 11:4467–4482

Dhingra MM, Sarma RH (1979) Nuclear magnetic resonance spectroscopy. Chemical shifts, coupling constants and molecular geometry. In: Sarma RH (ed) Stereodynamics of molecular systems. Pergamon, pp 3–38

Dickerson RE, Drew HR (1981) Structure of a B-DNA dodecamer II. Influence of base sequence on helix structure. J Mol Biol 149:761–786

Dickerson RE, Drew HR, Conner BN, Wing RM, Fratini AV, Kopka ML (1982) The anatomy of A-, B-, and Z-DNA. Science 216:475–485

Doornbos J, Charubala R, Pfleiderer W, Altona C (1983) Conformational analysis of the trinucleoside diphosphate 3′d(A2′-5′A2′-5′A). An NMR and CD study. Nucl Acids Res 11:4569–4582

Drew HR, Dickerson RE (1982) A new model for DNA containing A.T and I.C base pairs. EMBO J 6:663–667

Early TA, Kearns DR, Hillen W, Wells RD (1980) A 300 MHz and 600 MHz proton NMR study of a 12 base pair restriction fragment: Investigation of structure by relaxation measurements. Nucl Acids Res 8:5795–5812

Evans FE, Sarma RH (1974) The intramolecular conformation of adenosine 5′-monophosphate in aqueous solution as studied by fast Fourier transform ^1H- and ^1H-$\{^{31}P\}$ nuclear magnetic resonance spectroscopy. J Biol Chem 249:4754–4759
Feigon J, Denny WA, Leupin W, Kearns DR (1983a) Proton nuclear magnetic resonance investigation of the conformation and dynamics in the synthetic deoxyribonucleic acid decamers d(ATATCGATAT) and d(ATATGCATAT). Biochemistry 22:5930–5942
Feigon J, Leupin W, Denny WA, Kearns DR (1983b) Two-dimensional proton nuclear magnetic resonance investigation of the synthetic deoxyribonucleic acid decamer d(ATATCGATAT)$_2$. Biochemistry 22:5943–5951
Giessner-Prettre C, Pullman B (1970) Intermolecular nuclear shielding values for protons of purines and flavins. J Theor Biol 27:87–95
Hare DR, Wemmer DE, Chou SH, Drobny G, Reid BR (1983) Assignment of the nonexchangeable proton resonances of d(C-G-C-G-A-A-T-T-C-G-C-G) using two-dimensional nuclear magnetic resonance methods. J Mol Biol 171:319–336
Hilbers CW (1979) Hydrogen-bonded proton exchange and its effect on NMR spectra of nucleic acids. In: Shulman RG (ed) Biological applications of magnetic resonance. Academic Press, NewYork, pp 1–43
Holak TA, Borer PN, Levy GC, van Boom JH, Wang AHJ (1984) ^{31}P NMR analysis of the B to Z transition in double-stranded (dC-dG)$_3$ and (dC-dG)$_4$ in high-salt solution. Nucl Acids Res 12:4625–4635
James TL, Bendel P, James JL, Keepers JW, Kollman PA, Lapidot A, Murphy-Boesch J, Taylor JE (1983) Conformational flexibility of nucleic acids. In: Pullman B, Jortner J (eds) Nucleic acids: The vectors of life. D. Reidel, pp 155–167
Jeener J, Meier BH, Bachmann P, Ernst RR (1979) Investigation of exchange processes by two-dimensional NMR spectroscopy. J Chem Phys 71:4546–4553
Johnston PD, Redfield AG (1979) Proton FT NMR studies of tRNA structure and dynamics. In: Schimmel P, Abelson J, Soll D (eds) Transfer RNA: Structure properties and recognition. Cold Spring Harbor Laboratory, Cold Spring Harbor NY, pp 191–206
Kan LS, Cheng DM, Jayaraman K, Leutzinger EE, Miller PS, Ts'o POP (1982) Proton nuclear magnetic resonance study of a self-complementary decadeoxyribonucleotide, C-C-A-A-G-C-T-T-G-G. Biochemistry 21:6723–6732
Karplus M (1969) Proton spin coupling by Pi electrons. J Chem Phys 33:1842–1849
Kearns DR (1977) High resolution nuclear magnetic resonance studies of double helical polynucleotides. Annu Rev Biophys Bioeng 6:477–523
Kearns DR, Mirau PA, Assa-Munt N, Behling RW (1983) Two-dimensional NMR investigation of the structural properties of DNAs. In: Pullman B, Jortner J (eds) Nucleic acids: The Vectors of life. D Reidel, pp 113–125
Klug A, Jack A, Viswamitra MA, Kennard O, Shakked Z, Steitz TA (1979) A hypothesis on a specific sequence-dependent conformation of DNA and its relation to the binding of the lac-repressor protein. J Mol Biol 131:669–680
Kumar A, Ernst RR, Wuethrich K (1980) A two-dimensional nuclear Overhauser enhancement (2D NOE) experiment for the elucidation of complete proton-proton cross-relaxation networks in biological macromolecules. Biochem Biophys Res Commun 95:1–6
Lee HC, Tinocco I (1980) Conformation studies of 13 trinucleoside diphosphates by 360 MHz PMR spectroscopy. A bulged base conformation. I. Base protons and H1′ protons. Biophys Chem 11:283–294
Levitt M (1978) How many base pairs per turn does DNA have in solution and in chromatin? Some theoretical calculations. Proc Natl Acad Sci USA 75:640–644
Lown JW, Hanstock CC, Bleackly RC, Imbach JL, Rayner B, Vasseur JJ (1984) Synthesis, complete ^1H-assignments and conformation of the self-complementary hexadeoxyribonucleotide [d(CpGpApTpCpG)]$_2$ and its fragments by high field NMR. Nucl Acids Res 12:2519–2533
Mellema Jr, Jellema AK, Haasnoot CAG, van Boom JH, Altona C (1984) Conformational analysis of the single-helical DNA fragment d(T-A-A-T) in aqueous solution. The combined use of NMR proton chemical shifts and coupling constants obtained at 300 MHz and 500 MHz. Eur J Biochem 141:165–175

Mitra CK, Sarma MH, Sarma RH (1981) Left-handed deoxyribonucleic acid double helix in solution. Biochemistry 20:2036–2041

Noggle JH, Schirmer RE (1971) The nuclear Overhauser effect. Academic Press, New York

Pardi A, Walker R, Rapoport H, Wider G, Wuethrich K (1983) Sequential assignments for the ^{1}H- and ^{31}P atoms in the backbone of oligonucleotides by two-dimensional nuclear magnetic resonance. J Am Chem Soc 105:1652–1653

Patel DJ, Canuel LL, Pohl FM (1979) "Alternating B-DNA" conformation for the oligo(dG-dC) duplex in high-salt solution. Proc Natl Acad Sci USA 76:2508–2511

Patel DJ, Kozlowski SA, Nordheim A, Rich A (1982) Right-handed and left-handed DNA: Studies of B- and Z-DNA by using proton nuclear Overhauser effect and P NMR. Proc Natl Acad Sci USA 79:1413–1417

Patel DJ, Kozlowski SA, Marky LA, Rice JA, Broka C, Dallas J, Itakura K, Breslauer KJ (1982a) Structure, dynamics, and energetics of deoxyguanosine thymidine wobble base pair formation in the self-complementary d(CGTGAATTCGCG) duplex in solution. Biochemistry 21:437–444

Patel DJ, Kozlowski SA, Marky LA, Rice JA, Broka C, Itakura K, Breslauer KJ (1982b) Extra adenosine stacks into the self-complementary d(CGCAGAATTCGCG) duplex in solution. Biochemistry 21:445–451

Patel DJ, Kozlowski SA, Marky LA, Rice JA, Broka C, Itakura K, Breslauer KJ (1982c) Structure and energetics of a hexanucleotide duplex with stacked trinucleotide ends formed by sequence d(GAATTCGCG). Biochemistry 21:451–455

Patel DJ, Kozlowski SA, Marky LA, Broka C, Rice JA, Itakura K, Breslauer KJ (1982d) Premelting and melting transitions in the d(CGCGAATTCGCG) self-complementary duplex in solution. Biochemistry 21:428–436

Patel DJ, Kozlowski SA, Bhatt R (1983) Sequence dependence of base-pair stacking in right-handed DNA in solution: Proton nuclear Overhauser effect NMR measurements. Proc Natl Acad Sci USA 80:3908–3912

Reid DG, Salisbury SA, Williams DH (1983) A ^{1}H-NOE and CD study of the salt-concentration dependence of the structure of $d(G\text{-}C)_5$. Nucl Acids Res 11:3779–3793

Robillard GT, Reid BR (1979) Elucidation of nucleic acid structure by proton NMR. In: Shulman RG (ed) Biological applications of magnetic resonance. Academic Press, New York, pp 45–112

Sanderson MR, Mellema JR, van der Marel GA, Wille G, van Boom JH, Altona C (1983) Assignment of non-exchangeable base proton and H1′ resonances of a deoxyoctanucleoside heptaphosphate d(G-G-C*-C*-G-G-C-C) by using the nuclear Overhauser effect. Nucl Acids Res 11:3333–3346

Sarma RH, Mitra CK, Sarma MH (1981a) Structure of the DNA double helix in solution, crystals and theory. Coming of age of Z-DNA. In: Sarma RH (ed) Biomolecular stereodynamics. I. Adenine, pp 53–75

Sarma RH, Wagner BJ, Mitra CK (1981b) Propeller twisted adenine-thymine base pairs in the DNA double helix in solution. Solution spatial configuration of the AATT domain in the double helices of d-GGAATTCC and d-CGCGAATTCGCG. In: Sarma RH (ed) Biomolecular stereodynamics I. Adenine, pp 89–98

Scheek RM, Russo N, Boelens R, Kaptein R, van Boom JH (1983) Sequential resonance assignments in DNA. ^{1}H-NMR spectra by two-dimensional NOE spectroscopy. J Am Chem Soc 105:2914–2916

Scheek RM, Boelens R, Russo N, van Boom JH, Kaptein R (1984) Sequential resonance assignments in ^{1}H-NMR spectra of oligonucleotides by two-dimensional NMR spectroscopy. Biochemistry 23:1371–1376

Schweizer MP, Broom AD, Ts'o POP, Hollis DP (1968) Studies of inter- and intramolecular interaction in mononucleotides by proton magnetic resonance. J Am Chem Soc 90:1042–1055

Tran-Dinh S, Neumann JM, Huynh-Dinh T, Igolen J, Kan SK (1982) DNA fragment conformations. I. Methods for the assignment of DNA fragment proton resonances. ^{1}H-NMR spectra of dApTpGpT and dApCpApTpGpT. Org Mag Res 18:148–152

Ts'o POP, Kondo NS, Schweizer MP, Hollis DP (1969) Studies of the conformation and interaction in dinucleoside mono- and diphosphates by proton magnetic resonance. Biochemistry 8:997–1029
Weiss MA, Patel DJ, Sauer RT, Karplus M (1984a) ^{1}H-NMR study of the λ operator site O_L1: Assignment of the imino and adenine H2 resonances. Nucl Acids Res 12:4035–4047
Weiss MA, Patel DJ, Sauer RT, Karplus M (1984b) NMR Characterization of DNA: Assignment of major groove sugar protons the λ-phage operator size O_L1 by two-dimensional NOE and J-correlated spectra. J Am Chem Soc 106:4269–4270
Wemmer DE, Chou SH, Hare DR, Reid BR (1984) Sequence-specific recognition of DNA: Assignment of nonexchangeable proton resonances in the concensus Pribnow promoter DNA sequence by two-dimensional NMR. Biochemistry 23:2262–2268
Wing R, Drew H, Takano T, Broka C, Tanaka S, Itakura K, Dickerson RE (1980) Crystal structure analysis of a complete turn of B-DNA. Nature 287:755–758

NMR Methods for Determination of Intracellular pH

J. K. M. Roberts

1 Introduction

The importance of intracellular pH regulation is well recognized (for review, see Smith and Raven 1979; Roos and Boron 1981; Poole-Wilson 1978). Two aspects in particular are of interest currently. First there is the question of the extent to which, and the mechanisms by which, the pH in the various intracellular compartments is regulated despite the occurrence of processes – such as proton extrusion, leakage of H^+ between intracellular compartments, CO_2 fixation – that would otherwise change intracellular pH. Second, intracellular pH changes may mediate changes in the rates of certain biochemical reactions in cells, pH acting as a second messenger that transmits information from one part of a cell to another. A prerequisite for appreciation of the role that these two types of phenomenon play in metabolism are pH measurements of the various intracellular compartments. The difficulties of making such measurements has hindered progress toward this goal (Smith and Raven 1979). NMR methods for pH measurement in plant cells should improve this situation.

In this chapter the principles, practice, advantages and limitations of NMR methods for intracellular pH measurement are described. The ^{31}P-NMR method for cytoplasmic and vacuolar pH measurement is considered here in most detail, largely because this is the most widely used method, and in some respects the easiest to apply. Nonetheless, it is hoped that the general principles, as they apply to intracellular pH measurements using nuclei other than ^{31}P, will be noted. This is because techniques that employ ^{1}H, ^{13}C, ^{15}N, or ^{19}F signals as intracellular pH indicators, while at present having been used in only a very few studies, will undoubtedly offer the means to obtain important physiological information.

NMR spectroscopic techniques can be used to measure the pH of the cytoplasm and inside vacuoles. At present there is little or no evidence that pH inside organelles such as mitochondria, chloroplasts, Golgi bodies, and starch grains can be measured in intact tissues using NMR methods (see below). The principal advantages of NMR methods are that measurements of cytoplasmic and vacuolar pH can be made simultaneously, nondestructively, continuously (over periods from less than 1 min to many days) and in many cases noninvasively – i.e., without forcing foreign chemicals or objects into cells. Some measure of the degree of variation in pH in a particular type of compartment throughout the tissue may also be obtained. These pH measurements can be made on almost any plant tissue.

That a description of the factors limiting the accuracy of pH measurements by NMR comprises much of this chapter should not be taken to indicate that the method is less accurate or less useful than other methods for determination of intracellular pH. Rather, these factors are emphasized because they must be appreciated in order to apply the method correctly. It turns out, in fact, that theoretical and practical problems associated with the method, while limiting the number of decimal places to which intracellular pH measurements can be made, inhibits the ability of the method to answer physiological questions very little. Assessment of whether or not NMR is the best way to obtain pH measurements in a particular biological system involves comparison with other methods; these are made elsewhere (Roos and Boron 1981; Poole-Wilson 1978). A useful description of the relationships between concepts and measurements of pH is given by Bates (1981).

2 Some Principles

Intracellular pH measurements are made using NMR as follows. (a) The tissue or cells of interest are placed in an NMR (test) tube modified to permit perfusion of tissues, or oxygen bubbling of cell suspensions. The sample is placed in the spectrometer, which is set to observe the nucleus of interest (e.g., ^{31}P). (b) A spectrum is obtained, and (c) the positions of particular peaks (spectral lines) are measured, relative to a reference compound. ^{1}H and ^{31}P spectra of living material consist of peaks from naturally occurring, mobile metabolites that contain hydrogen or phosphorus – these isotopes being 100% abundant in nature. On the other hand, ^{13}C and ^{15}N are present at low percentage abundance in nature, while ^{19}F does not occur naturally in cells in combined form. Hence, for intracellular pH measurements using ^{13}C-, or ^{15}N-, or ^{19}F-NMR, cells must be provided with a labeled compound that can be taken up and accumulated to sufficient concentration (ca. millimolar). (d) The various peaks in the spectrum are then assigned, both to the particular type of chemical group giving rise to the signal, and to the location(s) of each chemical within the cells or tissues. Observation and assignment of a peak to a chemical group whose position in the spectrum ("chemical shift") is known to be pH-dependent sets the stage for pH measurement of the compartment in which the chemical group is located. (e) The measurement of pH is made by relating the observed chemical shift of the pH-sensitive peak to pH, using a titration curve of chemical shift versus pH obtained in vitro. The accuracy of the pH measurement is determined, first, by the accuracy with which the chemical shift of the particular line is determined and, second, by the validity of the titration curve used to relate chemical shift to pH. These constraints are considered below in more detail.

Magnetic resonance can be observed in populations of nuclei, such as ^{1}H, ^{13}C, ^{15}N, ^{19}F, and ^{31}P, that possess a permanent magnetic moment attributable to nuclear spin. When placed in an external magnetic field, these nuclei orient their magnetic moments either parallel with (ground state) or antiparallel to (excited state) the external magnetic field. At thermal equilibrium there is a tiny excess of nuclei in the ground state. This excess, or a fraction of it, can be raised to the excited state by absorption of radiofrequency (rf) electromagnetic radiation of frequency corresponding to the difference in energy between the two states. For example, when ^1H nuclei are placed in an external magnetic field of 100 kG only 1 in $\sim 10^6$ are seen to undergo magnetic resonance. It is for this reason that NMR is one of the most insensitive spectroscopic techniques.

Modern NMR experiments consist of first irradiating a population of nuclei in a sample with a pulse (or a series of pulses) of rf radiation, and then recording the signal that comes from the excited nuclei. This signal decays with time (and so is called a free induction decay, FID) as net movement of nuclear spins to the ground state occurs until thermal equilibrium is reached. The phenomenon of reestablishment of thermal equilibrium in a spin population is termed longitudinal relaxation; it usually takes place exponentially, with a time constant T_1. Fourier transformation (FT) of the FID (signal intensity as a function of time) gives a conventional spectrum (signal intensity as a function of frequency). The insensitivity of NMR spectrometers, together with the low concentrations of metabolites (other than H_2O) in living tissue, requires that signal averaging be performed, to obtain sufficient spectral signal relative to noise. Thus the usual in vivo FT-NMR experiment involves adding several to many thousand FID's obtained using the sequence: exciting pulse, acquisition, delay to allow for longitudinal relaxation, exciting pulse, etc. Commonly used time periods for the steps in this cycle are 10–

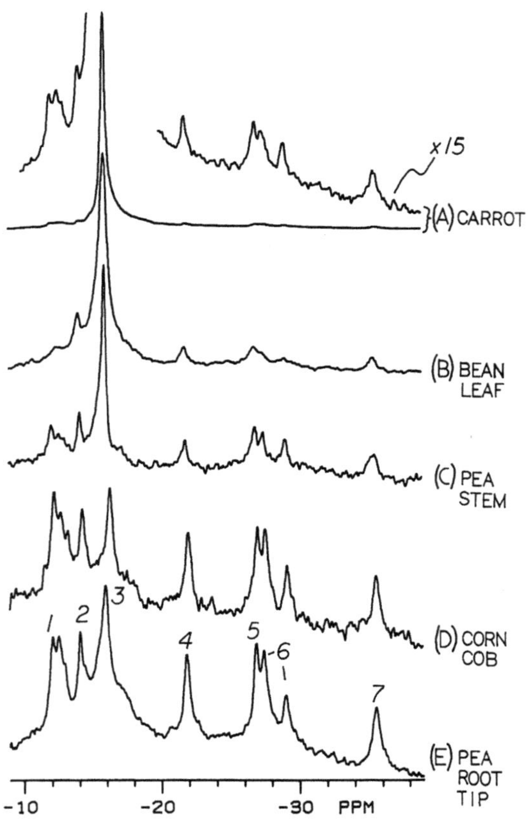

Fig. 1. ^{31}P-NMR spectra of various plant tissues. Tissues were perfused with oxygen-saturated 30 mM sucrose at 40 ml min^{-1}, except for the carrot storage tissue, which was treated with a stream of humidified oxygen, and the bean leaf, which was perfused with 0.1 mM $CaSO_4$. Spectra were obtained in modified Bruker HXS-360 operating at 145.7 MHz, over a period of 1 h, except for *A* and *B*, which were obtained over 4 h. Peak assignments: *1* glucose-6-phosphate; *2* cytoplasmic inorganic phosphate; *3* vacuolar inorganic phosphate; *4* γATP; *5* αATP; *6* uridine diphosphoglucose and nicotinamide adenine nucleotides; *7* βATP. (Roberts 1984)

30 µs for the exciting pulse, 0.2–1 s for acquisition, and 0.2–4 s for the delay before the next exciting pulse (T_1's for most nuclear spins in vivo are on the order of 1 s). The NMR spectrum is then obtained by FT of the sum of these FID's. This, then, briefly summarizes how NMR spectra are obtained. Further details of NMR theory, and its practice, can be found in the monographs by Farrar and Becker (1971), Shaw (1976) and Becker (1980).

NMR spectra of plants contain a number of Lorentzian-shaped absorption lines, some of which overlap (see Fig. 1). Only absorption lines from nuclei in freely mobile compounds (correlation time less than that of large proteins) are detected in conventional NMR experiments. Moreover, absorption lines from only one isotope (e.g., only ^1H, or only ^{31}P) are observed in a given NMR experiment. Nuclei of different isotopes absorb rf radiation at very different frequencies (separated by MHz) in a given magnetic field, whereas the various absorption lines from any given isotope are separated by no more than a few thousand Hz.

3 Determinants of Chemical Shift

The frequency of radiation required to excite a given population of nuclei is linearly proportional to the magnetic field strength they experience. The various absorption lines in an NMR spectrum arise from subpopulations of nuclei in the sample that are experiencing slightly different magnetic field strengths (so that the nuclei have different resonance frequencies). Environments differing in magnetic field strength can result from one, or a combination, of three effects.

1. Nuclei in different chemical groups are shielded from the external magnetic field to different extents by the electrons in the chemical group, and so experience magnetic fields that differ in strength. This phenomenon is called the "chemical shift". It is because the chemical shifts (i.e., resonance frequencies) of nuclei in some chemical groups are strongly pH-sensitive that intracellular pH measurements can be made using NMR spectroscopy. Such chemical groups can be used as pH indicators, provided the relationship between chemical shift and pH in vivo is known. The pH dependence of the chemical shift of an NMR line is due to changes in the distribution of electrons around a nucleus as the chemical group ionizes. As will be seen below, the time-scale of NMR is such that only the average ionization state is observed. For example, a ^{31}P-NMR spectrum of potassium phosphate, at its pK_a, consists of a single peak at a frequency mid-way between the positions of signals given by solutions of pure KH_2PO_4 and pure K_2HPO_4; i.e., instead of seeing a solution that contains equal amounts of $H_2PO_4^-$ and HPO_4^{2-}, NMR sees a solution consisting of $H_{1.5}PO_4^{1.5-}$. An exception to this rule is the equilibrium $CO_2+H_2O=H_2CO_3=H^++HCO_3^-$, which even in vivo (see Roberts and Jardetzky 1981) appears to be established sufficiently slowly for separate CO_2 and HCO_3^- ^{13}C-NMR lines to be observed in cells provided with ^{13}C-labeled bicarbonate.

2. The bulk magnetic susceptibility, κ, (see Becker 1980) of the medium about the nuclei under observation also influences the magnetic field experienced by the

nuclei, and so their chemical shift. The bulk magnetic susceptibility in any region of a sample is determined by the chemical composition, sample shape, and sample orientation with respect to the magnetic field. Bulk magnetic susceptibility may therefore be nonuniform throughout a sample – particularly in samples as chemically and physically anisotropic as living tissues. The effect of κ on chemical shift can be significant relative to the effects of changes in chemical state (e.g., ionization state, metal ion binding, proximity of other groups in a molecule). Bulk magnetic susceptibility effects are important in regard to the validity of chemical shift measurements which, in turn, determines in part the accuracy of intracellular pH measurements; this is discussed below.

3. The resonance frequencies of nuclei observed in a sample are further dependent upon the degree of inhomogeneity in the external magnetic field throughout the sample volume. Inhomogeneity may arise either from intrinsic imperfections in the magnet of a spectrometer (poor construction and/or "shims"), or from field distortions due to inhomogeneity in sample composition. The former contribution is usually insignificant in NMR studies of living biological material. Rather, the lumps and, in some cases, air spaces that constitute most in vivo NMR samples result in field gradients throughout the sample (cf. Waterton et al. 1983). The simplest effect these gradients have on the observed spectrum

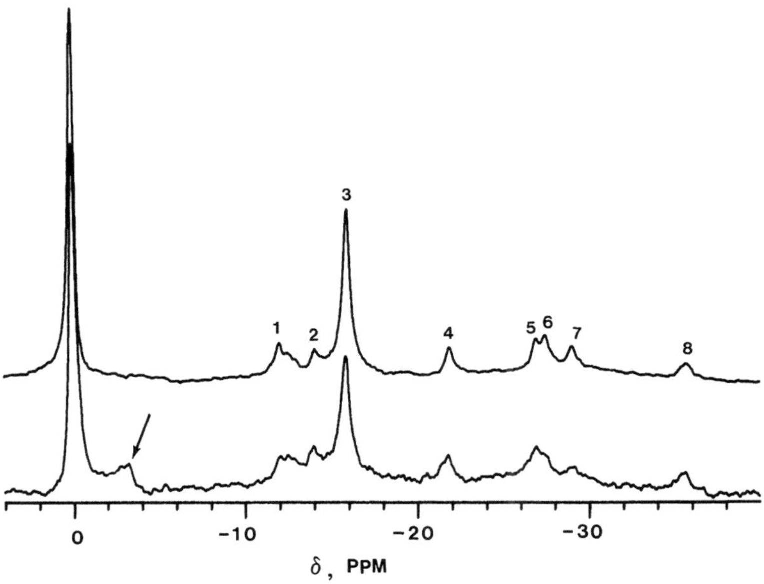

Fig. 2. Effect of air bubbles on ^{31}P-NMR of perfused, etiolated pea stems. The upper spectrum is of a sample in which all but the smallest air bubbles (<0.5 mm; cf. Ugurbil et al. 1979) were absent. The lower spectrum is of a similar sample, except that some large air bubbles were present between the cylindrical pieces of tissue. The greater magnetic field inhomogeneity, caused by the air bubbles, can be seen to result in broader spectral lines, and the appearance of a spurious resonance at ~ -3 ppm (see text). Assignments: *1* glucose-6-phosphate; *2* cytoplasmic inorganic phosphate; *3* vacuolar inorganic phosphate; *4* γATP; *5* αATP; *6, 7* uridine diphosphoglucose and nicotinamide adenine nucleotides; *8* βATP. The peak at 0 ppm is due to methylene diphosphonate, included in the sample volume in a coaxial capillary

is to broaden lines, and so decrease spectral resolution and the spectral signal relative to noise; this decreases the ease with which changes in the chemical shift of a particular line can be observed. Inhomogenous samples can also lead to the creation of discrete "pockets" within the sample differing in magnetic field strength, even though the sample is placed in a (formerly) homogenenous field. Such pockets will cause a specific chemical group in a specific chemical environment (and in a medium of the same bulk magnetic susceptibility) to give two or more spectral lines, instead of the single line observed when a homogenous magnetic field exists. Figure 2 illustrates the effects of air bubbles on ^{31}P-NMR spectra of perfused etiolated pea stems.

4 Measurement of Chemical Shifts

As noted above, the precise frequency at which populations of nuclei resonate is proportional to the magnetic field strength. Consequently, in order to compare chemical shifts of lines in NMR spectra obtained using spectrometers differing in magnetic field-strength, it is usual to report chemical shifts, given the symbol δ, in the dimensionless unit of parts per million (ppm), defined as

$$\delta = \frac{\nu_s - \nu_r}{\nu_r} \times 10^6;$$

where ν_s and ν_r are the resonance frequencies of the particular line of interest and the reference line, respectively. Such a relative measurement is entirely analogous to the δ^{13}C measurement (in the dimensionless units of parts per thousand) used in carbon mass spectrometry. Thus, a reference line is necessary for measurement of spectral line positions. The requirements a spectral line must meet to be useful for chemical shift determinations in biological NMR studies are (1) the line must be sufficiently intense, so that its apparent position is not influenced by spectral noise; (2) the line must be narrow, so that inaccuracies in the chemical shift measurements are not increased because of uncertainties as to the precise frequency of the maximum (center) of the reference line; and (3) the position (and intensity) of the line must be unaffected by changes in physiological state or experimental conditions.

Two kinds of reference compound should be distinguished. One kind includes substances inside the cells under investigation, and is called an internal reference. Examples include creatine phosphate in ^{31}P-NMR studies of muscle (Gadian 1982), and the 1-C glucose resonance in ^{13}C-NMR spectra of cells fed [1-^{13}C]-glucose (Ugurbil et al. 1978). The principal advantage of an internal reference is that chemical shift measurements may contain smaller (or no) contributions from variations in bulk magnetic susceptibility in the sample volume – for the "reference" and "sample" compounds are in the same (or similar) milieu. However, this method is limited in that in vivo NMR spectra do not always contain strong and "inert" resonances. It is also difficult to show that the resonance frequency of a potential internal reference line is independent of the physiological condition of the material under study, and the experimental conditions (e.g., temperature) – or to correct for such effects. Consequently, it is common to use external references: compounds contained in thin capillaries and place next to or amongst the biological material in the sample volume. For example, for in vivo ^{31}P-NMR studies, methylene diphosphonate is a useful reference compound – resonating at a frequency far from the resonance frequencies of phosphates observed in vivo (Fig. 2). The problem inherent to chemical shift measurements using external reference compounds stem from bulk magnetic susceptibility effects, considered above. The contribution of bulk magnetic susceptibility is difficult to separate cleanly from other determinants of observed chemical shifts, such as pH. This is one factor that must be considered when estimating the accuracy of intracellular pH measurements.

5 Estimating Intracellular pH from Chemical Shifts: Titration Curves

In 1973 Moon and Richards described how the pH inside erythrocytes could be estimated from the ^{31}P chemical shifts of 2,3-diphosphoglycerate and inorganic phosphate observed in spectra of whole rabbit blood, or packed erythrocytes. They determined in detail the relationship between the pH in hemolysates, determined using a pH-sensitive electrode, and the ^{31}P-NMR chemical shifts of these two phosphates. The relationships were expressed as titration curves: plots of δ versus pH. They observed ^{31}P-NMR signals from blood that were assigned to 2,3-diphosphoglycerate and inorganic phosphate, and determined their chemical shifts. Intracellular pH was then estimated by reading off the hemolysate pH that corresponded to the chemical shifts of the various phosphates observed in vivo. All intracellular pH measurements using NMR spectroscopy involve analogous steps: generating titration curves, measuring in vivo chemical shifts, and relating the two. The following describes the approaches used to relate chemical shifts to true intracellular pH via titration curves. Later, other aspects in pH measurements using in vivo NMR are considered.

As noted by Moon and Richards (1973), a central obstacle to accurate determination of intracellular pH by NMR exists because factors other than pH can also significantly influence NMR chemical shifts, i.e., the precise forms that δ versus pH titration curves take. The problem is illustrated by Fig. 3 in which the behavior of the ^{31}P chemical shift of inorganic phosphate as a function of pH is compared in different solutions: 5 mM potassium phosphate, homogenated maize root tips, and homogenated carrot storage tissue. The three titration curves are clearly different. The differences can be rationalized in terms of the effect of factors such as the ionic strength and the concentration of free divalent ions (Roberts et al. 1981 b). Two types of effect may be distinguished. First there are simple effects of ions on the pK_a of the titrating group, expected as ions compete with protons for the titrating group. This effect results in a decrease in the pK_a of the group as the concentration of the ions increases. The reciprocal relationship

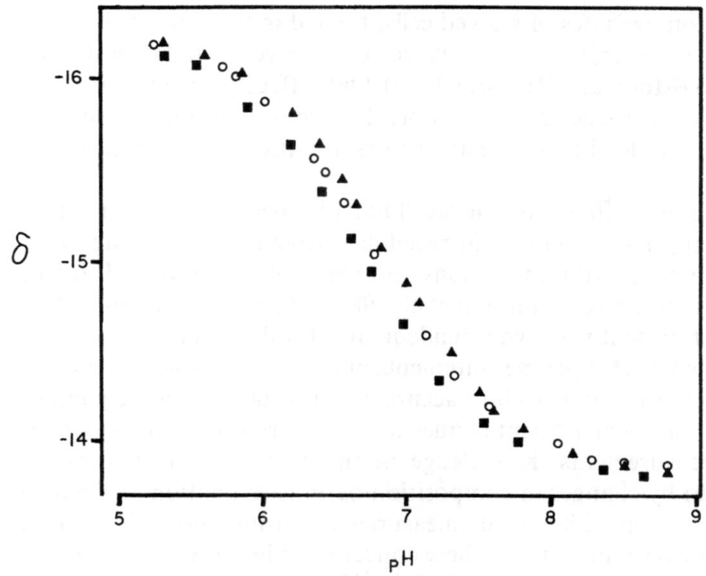

Fig. 3. Titration curves of inorganic phosphate in solutions differing in composition. The graph plots the chemical shift, δ. of the inorganic phosphate ^{31}P resonance (observed at 145.7 MHz and relative to 0.5 M methylene diphosphonate contained in a capillary in the sample tube) as a function of pH, measured using a combination pH electrode. The different symbols refer to data obtained with either 5 mM potassium phosphate (▲), or the phosphate in an undiluted homogenate of 2 mm long maize root tips (centrifuged at 10,000 *g* for 30 min) (○), or the phosphate in an undiluted homogenate of carrot storage tissue (filtered through miracloth) (■)

between pK_a and ionic strength is stronger for ions of greater valency, as might be expected. Second are more specific effects of ions on the chemical shifts of titratable group, i.e., effects that cannot be attributed to changes in pK_a. Such effects, perhaps, can be rationalized by noting that the distribution of electrons around ^{31}P in a HPO_4^{2-} ion, for example, might well be distorted on binding a magnesium ion.

It is these effects of ions that render ATP ^{31}P resonances useless for intracellular pH measurements in the vicinity of pH 7. Thus, the γATP line of pure ATP changes greatly over the range pH 5.5 to 8. However, in the presence of physiological concentrations of magnesium ion [ATP in vivo is almost invariably present as MgATP (Gadian 1982; Wray et al. 1983)], and at physiological ionic strength, the γATP line becomes much less sensitive to pH over the range normally encountered in vivo. An exception is pH measurements of the internal matrix of isolated chromaffin granules, which have an unusual composition (see Roberts and Jardetzky 1981).

Regardless of how well differences in titration curves of δ versus pH are explained, the effects just described clearly point to a need for caution in the use of titration curves for intracellular pH measurements by NMR. The problem is not so great for pH measurements of cells, such as erythrocytes, that contain only one intracellular compartment. A titration curve derived from the particular compound in undiluted homogenates of packed cells, treated in the same way as the cells used for in vivo experiments, can be expected to be very similar to that of the compound in vivo (Moon and Richards 1973). Only effects of changes in composition on titration curves, due to chemical breakdown or addition of acid and base during titration, will lead to systematic errors in intracellular pH measurements.

The situation is more difficult for intracellular pH measurements in other, more typical cells that possess complex intracellular structures. The existence of large gradients in the concentrations of ions and metabolites between different compartments is well known (e.g., Pitman et al. 1981). Methods for quantitating these gradients are at present somewhat inadequate. It follows that use of titration curves of homogenates for pH measurements of particular intracellular compartments is not a definitive approach to accuracy. Uncertainties in the composition of the intracellular compartment in question will therefore contribute to inaccuracies in pH measurements. Knowledge of the extent to which titration curves can be affected by changes in composition permits estimation of the possible systematic errors in intracellular pH measurements using NMR (Roberts et al. 1981 b). The practical significance of these concerns is illustrated by the observation that the chemical shift of the cytoplasmic ^{31}P inorganic phosphate line is the same in both normoxic carrot storage tissue and maize root tips – even though the titration curves for inorganic phosphate in homogenates of these tissue are different (Fig. 3). This observation raises the question: is cytoplasmic pH different in these two tissues, or are they in fact the same? – the difference in titration curves being due to differences in the chemical composition in compartments other than cytoplasm. Uncertainties such as these can to some extent be alleviated in NMR experiments in which more than one intracellular pH indicator is available, as is possible in ^{31}P-NMR measurements of cytoplasmic pH in glucose-fed plant tissues (Roberts et al. 1980, 1982). Because the titration curves of glucose-6-phosphate and inorganic phosphate are altered to different extents by changes in ionic strength and divalent ion strength (Roberts et al. 1981 b), the responses of the ^{31}P chemical shifts of these phosphates to changes in intracellular pH can be used to validate the in vitro titration curves used to estimate intracellular pH. If the frequency difference between the glucose-6-phosphate and the inorganic phosphate lines is accurately described by the titration curves as intracellular pH changes, the intracellular pH measurements are likely to be very accurate. Such a rigorous validation has not been made, perhaps in part because the chemical shift measurements required are difficult to make with sufficient accuracy.

While absolute intracellular pH measurements cannot be made rigorously without some uncertainties, the problems associated with measurement of intracellular pH changes are much less acute. Here the only questions to be answered are, first, is the potential NMR pH indicator sensitive to pH in vivo? and, second, what is the slope of the titration curve of the compound over the pH ranges to be measured? The first question can be answered simply by showing that the NMR signal shifts when intracellular pH changes. For example, the cytoplasmic inorganic phosphate ^{31}P signal in maize root tips moves in a manner consistent with alkalization when they are treated with ammonia (Roberts et al. 1980, 1981 a, 1982), and in a manner consistent with acidification when they are treated with acetic acid (Roberts et al. 1984a), carbon dioxide (Roberts et al. 1984c), or under hypoxia when lactic acid accumulates in the cytoplasm (Roberts et al. 1982, 1984a).

The second question can be answered by comparing the slopes of ^{31}P titration curves of phosphates in solutions differing in composition (Roberts et al. 1981b); these slopes are quite similar for solutions varying in composition over a physiologically possible range (Roberts et al. 1981b; 1982). Hence, while absolute cytoplasmic pH measurements in maize root tips cannot be shown definitively to be accurate to better than as much as ± 0.2 pH units, changes in pH can be measured to better than 0.1 pH unit (Roberts et al. 1982). Thus, intracellular pH regulation and intracellular pH changes can be observed and quantitated quite well by in vivo NMR. Such measurements are perhaps the most relevant for physiological studies; the value of absolute intracellular pH measurements accurate to better than one decimal place, were they possible, is not clear.

It should be noted that ^{31}P-NMR measurements of intracellular pH in *Escherichia coli* (Navon et al. 1977), *Neurospora* (Legerton et al. 1983), myocardium (Poole-Wilson 1978; Hollis and Nunnally 1980) and sea urchin eggs (Nuccitelli et al. 1981) have been found to agree well with pH measurements using other techniques. However, since all methods of pH measurement rely heavily on operationally defined calibration criteria (particularly when applied to measurement of hydrogen ion activity in solutions as complex, nonideal, heterogeneous, variable and ill-defined as intracellular milieux), this finding cannot be used to validate absolute intracellular pH measurements made by one or any method. All intracellular pH measurements suffer from similar problems of absolute calibration, in addition to other limitations usually specific to a particular method. The particular advantage specific to NMR methods of intracellular pH measurement is that pH regulation and pH changes can be quite accurately monitored continuously in a single sample, noninvasively; also, any variation in the pH's among the cells in the sample can also be followed.

6 Other Factors Contributing to Potential Errors in Intracellular pH Measurements Using NMR

In addition to inaccuracies in pH measurements due to uncertainties as to the correct titration curve to use, as just described, other factors also contribute errors. Systematic errors in chemical shift measurements of in vivo resonances, as they relate to chemical shifts in titration curves, will arise when an external reference compound is used, if the bulk magnetic susceptibility of compounds in vivo differs from that of the solutions used for titration. This subject is discussed above. Corrections for magnetic susceptibility effects are difficult to make, beyond saying that errors are most likely no greater than a few tenths of a ppm. Again, it is the accuracy of absolute intracellular pH measurements that is most affected by this uncertainty. Measurements of intracellular pH changes are not rendered so uncertain, because the slopes of titration curves do not change very much over such small variations in chemical shift. The ability to observe qualitatively whether or not intracellular pH changes under particular physiological conditions remains regardless of these concerns, provided it can be demonstrated that the NMR signal being used is pH-sensitive in vivo.

Errors in measurement of chemical shifts of spectral peaks, resulting from either poor spectral signal relative to noise, or broad (or overlapping) spectral lines, or both, necessarily introduce errors in both absolute and relative pH measurements. While this may be in some respects a trivial point, it can be a significant determinant of the accuracy of intracellular pH measurements, for two reasons. First, the ratio of signal to noise in spectra of cells and tissues is often low, particularly in studies in which rapid effects are under investigation (allowing for less signal averaging). This is because of the insensitivity of the method and the low concentrations of metabolites in vivo. Second, line-widths of resonances in spectra of cells and tissues are large relative to the amount the resonances shift with changing pH. For example, the inorganic phosphate ^{31}P-NMR resonances from cytoplasm and vacuoles in plant tissues (e.g., Figs. 1 and 2) overlap to some extent – even though their chemical shifts are on opposite ends of the ^{31}P inorganic phosphate titration curve (Fig. 3). Thus, although it is possible to measure "the" chemical shift of such lines with vanishingly small statistical error, such precise measurements belie the fact that these NMR signals suggest, to a certain extent, that a range of pH's in intracellular compartments exists. Such considerations lead to a further aspect of intracellular pH measurements using NMR techniques.

7 Observation of pH Heterogeneity in Cell Populations and Tissues by NMR

If the compound responsible for the pH-dependent NMR signal occurs at sufficient concentrations in regions of the sample differing in pH, more than one NMR line will be observed. However, the fact that in vivo resonances are broad relative to the range over which chemical shifts change with changing pH puts a limit on the number of separate pH "compartments" that can be observed. Hence, most often pH variation is recognized by the existence of broader lines, or "shoulders" on lines. Figure 4 illustrates this attribute of the NMR method. Extreme cytoplasmic acidosis occurs in maize root tips during prolonged hypoxia, in part due to leakage of acid from the vacuole – vacuolar alkalinization occurring concomitantly (Roberts et al. 1984c). The ^{31}P-NMR spectra in Fig. 4 indicate that a much wider range of cytoplasmic pH's exists in the hypoxic root tips, compared to the normoxic tissue, for the cytoplasmic inorganic phosphate line is much broader. In contrast, heterogeneity of cytoplasmic pH's does not increase significantly during shorter periods of hypoxia (Roberts et al. 1984a), or under conditions of less complete inhibition of oxidative phosphorylation (Fig. 5).

Inferences of pH variation in the sample using NMR data must be made with caution. Other factors contribute to in vivo spectral line widths, and these must

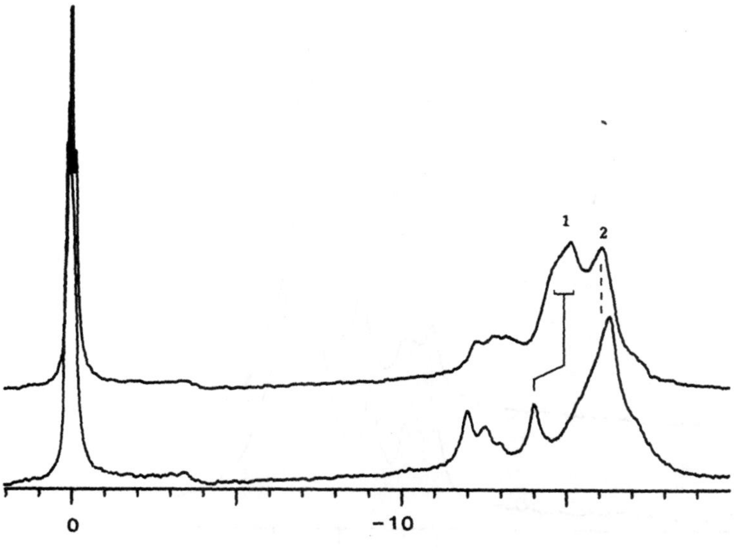

Fig. 4. Observation of heterogeneity in cytoplasmic pH in maize root tips by ^{31}P-NMR. The lower spectrum shows sugar and inorganic phosphate resonances from normoxic maize root tips. The upper spectrum is of the same sample, after ca. 20 h of hypoxia. Peak *1*, due to cytoplasmic inorganic phosphate, can be seen to be greatly broadened in prolonged hypoxia (compared to the normoxia), indicating that the cytoplasms of different cells in the sample (ca. 2 g total) are at different acidic pHs. The shift in the vacuolar inorganic resonance (peak *2*) indicates that alkalinization of the vacuole has occurred concomitant with cytoplasmic acidosis. The peak at 0 ppm is due to methylene diphosphonate, used as an external reference

be accounted for before conclusions can be drawn. One factor, magnetic field inhomogeneity (discussed above), can be significant enough in some samples that it is not possible to observe heterogeneity in pH. Broad lines also can occur as a result of the presence of paramagnetic ions, such as manganous ions. The presence of paramagnetic ions in vacuoles of maize root tips appears to account in part for the broadness of the vacuolar ^{31}P inorganic phosphate resonance (Roberts et al. 1984b), rather than variations in pH between different vacuoles, as we previously speculated (Roberts et al. 1982). This topic is reviewed in more detail by Roberts and Jardetzky (1981).

A separate consideration concerning interpretation of NMR spectra is heterogeneity that can result from inadequacies in the maintenance of a defined physiological state throughout the sample volume. For example, broad ^{31}P-NMR signals from cytoplasmic inorganic phosphate are observed when insufficient oxygen is available to the plant tissue under study (Roberts, unpublished observations). This can be attributed to the more hypoxic cells in the sample having a lower cytoplasmic pH than the less hypoxic cells. Treatments which stimulate respiration can be expected to exaggerate the degree of heterogeneity in cytoplasmic pH. The need to use high cell densities, often with very active plant tissues

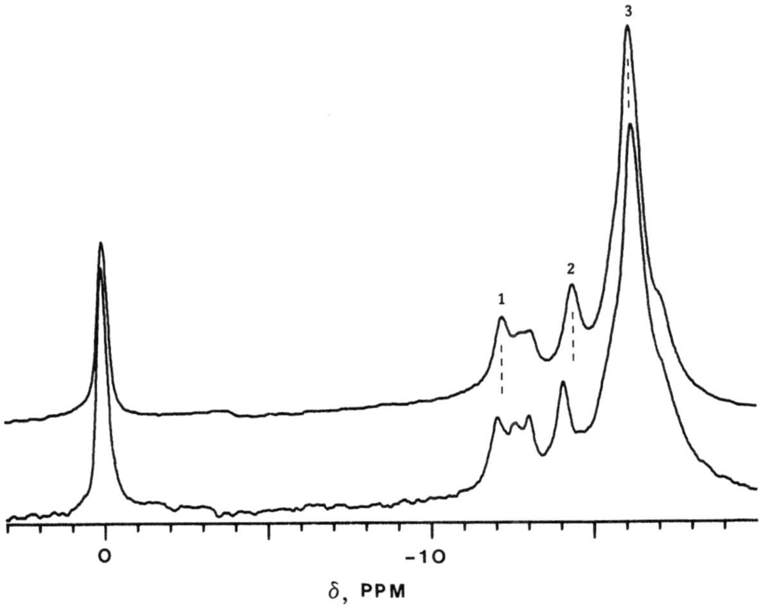

Fig. 5. Observation of cytoplasmic pH changes without a change in the degree of heterogeneity in pH in maize root tips. ^{31}P-NMR spectrum of oxygenated glucose-fed tissue treated with (*upper*) or without (*lower*) 1 mM salicylhydroxamic acid plus 100 μM KCN (pH 7). The shifts to the right of peaks *1* (glucose-6-phosphate) and *2* (cytoplasmic inorganic phosphate) indicate cytoplasmic acidification of ca. 0.3 pH units; vacuolar pH (given by the position of peak *3*) apparently does not change, although the chemical shift of inorganic phosphate is not very pH-sensitive at such low pH's (cf. Fig. 3). The peak at 0 ppm is due to methylene diphosphonate, used as an external reference

(in order to maximize spectral signal relative to noise) requires that special measures be taken to ensure that all the cells in the sample are maintained in the same physiological state. For active tissues this is achieved by perfusing (or rather, in animal jargon, "superfusing") the tissue with oxygen-saturated media at flow rates 40 ml min^{-1} (Roberts et al. 1982). This subject, the importance of which cannot be underemphasized, is considered in more detail in Chapter 2 of this volume "Determination of the energy status of plant cells using ^{31}P nuclear magnetic resonance". For NMR studies of cell suspensions, bubbling with oxygen gas suffices to met the respiratory needs of the cells (Martin et al. 1982).

8 NMR Signals That Have Been Used as Intracellular pH Indicators

Table 1 lists the various chemical groups that have been used to measure intracellular pH. Most intracellular pH measurements, in both plant and other biological systems, have used ^{31}P-NMR signals from naturally occurring inorganic or sugar

Table 1. NMR signals that have been used as intracellular pH indicators. Examples using animal or microbial cells are cited only when an example using plants is not available

Nucleus	pH indicator	Precursor supplied	Compartment studied	Cell-type	Reference
^{1}H	Histidine from Haemoglobin	None	Cytoplasm	Erythrocyte	Brown et al. (1977)
^{1}H	Organic acids	None	Intravesicle	Artificial	Cramer and Prestegard (1977) (cf. Ogino et al. 1980)
^{13}C	Malic acid	$^{13}CO_2$	Vacuole	CAM leaf	Stidham et al. (1983)
^{15}N	Amino acids	$^{15}NH_4Cl$ or [^{15}N]-histidine	Cytoplasm and vacuole	*Neurospora*	Legerton et al. (1983)
^{19}F	Fluorinated amino acid	Methyl ester of acid	Cell	Lymphocyte	Deutsch et al. (1982)
^{31}P	Inorganic phosphate	None	Cytoplasm and vacuole	Various	See Roberts (1984)
^{31}P	Glucose-6-phosphate	(Increased by external sugar)	Cytoplasm	Root tips	Roberts et al. (1980)
^{31}P	2-deoxyglucose-6-phosphate	2-deoxyglucose	Cytoplasm	Root tips	Roberts et al. (1980)

phosphates. Such measurements are technically easier to make in comparison to measurements using ^{13}C-, ^{15}N-, or ^{19}F-NMR. The latter approaches require selection of a suitable labeled compound, perhaps a precursor, that the tissue under study will take up and accumulate in sufficient quantities to be observable by NMR. Necessarily, protocols successful in one system are usually not generally applicable, because of metabolic differences between particular tissues types. Also, chemicals enriched in ^{13}C or ^{15}N, or analogs labeled with ^{19}F, can be expensive or difficult to obtain. Despite three impediments, there is little doubt that ^{13}C, ^{15}N, and ^{19}F-NMR methods will greatly extend the range of biological systems in which intracellular pH can be measured by NMR.

Two factors limit application of ^{1}H-NMR to intracellular pH measurement; no intracellular pH measurement in a plant using ^{1}H-NMR has been reported. First the frequency range over which proton chemical shifts occur is much smaller than in other nuclei. Hence, peaks in ^{1}H-NMR spectra are bunched close together, and the amount resonances shift with pH tends to be relatively little. These properties, coupled with the broad lines encountered in NMR spectra of tissues, usually mean that spectral resolution in ^{1}H-NMR in vivo is prohibitively poor. Only in cell suspensions are lines sufficiently narrow to permit resolution of the plethora of compounds giving ^{1}H signals (see, for example, Ogino et al. 1980). The second difficulty encountered during in vivo ^{1}H-NMR investigations is the problem of dynamic range: detection of signals from compounds present at approximately millimolar concentrations in the presence of a signal from water, present at ca. 110 M. Special techniques have been developed to avoid or limit this problem, and so increase the effective sensitivity of in vivo ^{1}H-NMR (see, for example, Hore 1983). Particular advantages of ^{1}H-NMR are its greater sensitivity compared to other nuclei, and the large number of ^{1}H-metabolites with titratable groups that exist at high concentrations in cells.

It is important, when using externally provided compounds to generate intracellular pH indicators – as in ^{13}C-, ^{15}N-, and ^{19}F-NMR – to show that the presence of these compounds does not perturb the experimental system under investigation. Such concerns are perhaps less acute when a substrate, normally present under natural conditions, is simply enriched with ^{13}C or ^{15}N; it is only necessary to ensure that excessive concentrations are not employed. However, when substances that do not naturally occur are used, it is important to show that the treatments employed are nonperturbing. This need is heightened by the requirement that these compounds or chemical groups accumulate to significant concentrations (millimolar) in order to be easily observed by NMR. An example where side effects are evident is the use of 2-deoxyglucose to measure cytoplasmic pH by ^{31}P-NMR (see Table 1). The principle of the method is that 2-deoxyglucose is readily taken up by cells, and phosphorylated by hexokinase in the cytoplasm; 2-deoxyglucose-6-phosphate accumulates in the cytoplasm, because its conversion to other metabolites is slow, and is observed by ^{31}P-NMR. This sequence of events is illustrated in Fig. 6. Figure 6 also shows that treatment with 2-deoxyglucose also results in a fall in cytoplasmic inorganic phosphate and ATP levels, presumably because cytoplasmic phosphate is sequestered as 2-deoxyglucose-6-phosphate accumulates, and because glycolysis is inhibited. While the concentrations of these metabolites eventually recover to near normal values after the

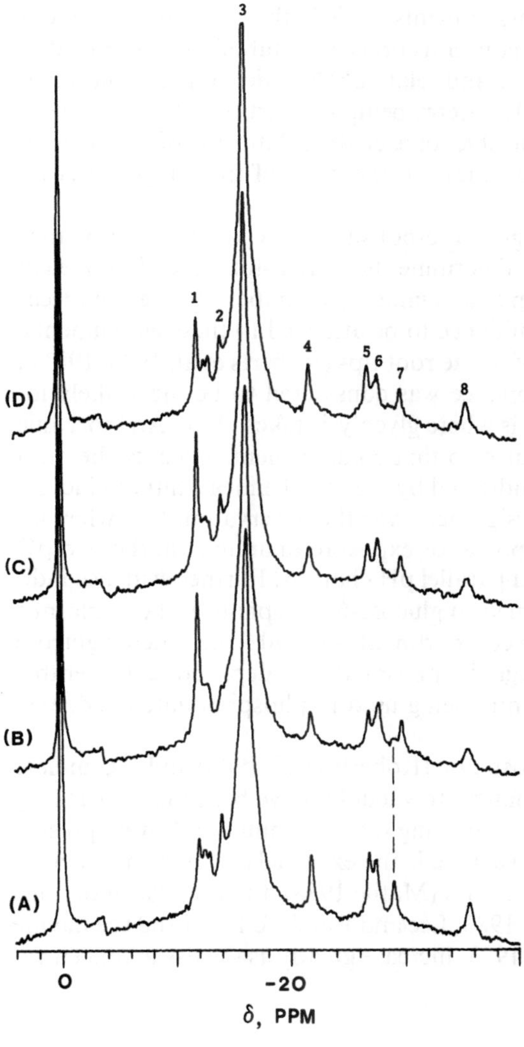

Fig. 6. Measurement of cytoplasmic pH by ^{31}P-NMR using 2-deoxyglucose. Spectra are of the same sample, taken sequentially, perfused first (*A*) with oxygenated 0.1 mM $CaSO_4$; then (*B*) with 50 mM 2-deoxyglucose/0.1 mM $CaSO_4$/oxygen saturated (acquired during the last hour of a 90-min treatment); then (*C*) as in *A* 3–4 h after the 2-deoxyglucose treatment; then (*D*) as in *A* 6–7 h after the 2-deoxyglucose treatment. Peak assignments: *1* glucose-6-phosphate and, after *A*, 2-deoxyglucose-6-phosphate; *2* cytoplasmic inorganic phosphate; *3* vacuolar inorganic phosphate; *4* γATP; *5* αATP; *6, 7* uridine diphosphoglucose and, after *A*, uridine diphospho-2-deoxyglucose; *8* βATP. Note that the chemical shift of peak *7* is shifted in spectra *B–D*, compared to *A*; this is attributed to formation of uridine diphospho-2-deoxyglucose, the phosphorus nucleus closest to the hexose apparently being sensitive to the presence or absence of oxygen at the C2 position. The peak at 0 ppm is due to methylene diphosphonate, used as an external reference

supply of 2-deoxyglucose is removed, this occurs only after a significant proportion of the 2-deoxyglucose-6-phosphate has been degraded. Hence the usefulness of this method is diminished.

9 Assignment of Resonances

Valid assignment of particular in vivo resonances is a prerequisite to intracellular pH measurements using NMR. Initially, resonances must be assigned to a particular chemical group. The first test applied here is to compare in vivo chemical shifts with chemical shifts of known metabolites under physiological conditions (pH, ionic strength, etc.). Tentative assignments made in this way can be checked by seeing if the absolute and relative concentrations of metabolites determined in tissue extracts conform to the absolute and relative intensities of particular resonances (the intensity of lines in NMR spectra being proportional to the concentrations of the chemical groups responsible for each line). Titration of extracts can also be used to aid in assignment of lines, on the basis of pK_a (Ugurbil et al. 1979).

Once the nature of the chemical group responsible for a particular resonance is determined, it is then necessary to determine the intracellular location of each chemical group. As with the first step in assignment, a number of different pieces of evidence may be required for confidence to be attached to these assignments. For example, in ^{31}P-NMR studies of maize root tips (Roberts et al. 1980, 1981 a, 1982), the glucose-6-phosphate resonance was considered to be most likely located in the cytoplasm, because that is where glycolysis takes place. An inorganic phosphate resonance was also assigned to that compartment, since its chemical shift indicated a pH similar to that indicated by the ^{31}P chemical shift of glucose-6-phosphate. Consistent with this assignment was the observation that when intracellular pH was perturbed by hypoxia, or exposure to ammonia, the two pH indicators suggested very similar and parallel pH changes. Further evidence indicating that the ^{31}P resonances assigned to glucose-6-phosphate and cytoplasmic inorganic phosphate are in the same compartment is found from 2-deoxyglucose experiments, for cytoplasmic inorganic phosphate is depleted considerably (Fig. 6, Roberts et al. 1980), apparently being used to phosphorylate the deoxyglucose (via ATP).

Assignment of ^{31}P inorganic phosphate (Roberts et al. 1980) and ^{13}C malate carboxyl (Stidham et al. 1983) resonances to vacuoles have been made primarily on the basis of chemical shifts, these indicating that the compounds were present in an acidic environment. Also, the vacuole is the expected location for metabolites present in large ("storage") quantities (Matile 1978); the vacuolar inorganic phosphate ^{31}P signal (Rebeille et al. 1983; Lee and Ratcliffe 1983) and the malate carboxyl ^{13}C signals (Stidham et al. 1983) increase greatly as the tissue concentrations of these compounds increase.

10 Can the pH in Compartments Other Than Cytoplasm and Vacuole be Measured in Vivo by NMR Methods?

While there is compelling evidence that cytoplasmic and vacuolar pH can be monitored by a variety of NMR approaches, there have been no reports of pH measurements of other intracellular compartments using in vivo NMR. This is so despite the fact that a number of titratable compounds exist in other intracellular compartments (such as amino and organic acids in mitochondria), and the fact that such compartments are expected to be at pH's different from the pH's of cytoplasm and vacuole. One possible reason for the absence of such measurements in plants is that compartments other than cytoplasm and vacuole usually occupy only a small proportion of the total tissue volume. This leads to very low concentrations of potential pH indicators, which are therefore not detectable by NMR. Difficulties in maintaining conditions in the spectrometer suitable for development of significant proton gradients between different intracellular compartments may also hinder observation of NMR signals from compartments such as the chloroplast (Waterton et al. 1983). The difficulties of assigning an extravacuolar inorganic phosphate signal to a specific compartment were also apparent in ^{31}P-NMR studies of photosynthetic cells and protoplasts (Foyer et al. 1982). Hence, it appears that the potential of NMR methods for pH measurement in compartments other than cytoplasm and vacuole remains to be determined. However, results relating to the measurement of mitochondrial pH by ^{31}P-NMR are possibly relevant to this question.

The proportion of tissue volume occupied by mitochondria in animals is considerable; in heart muscle it is as much as 40% (Bishop 1980). Despite this, in the considerable number of ^{31}P-NMR studies on isolated hearts (e.g., Matthews et al. 1981), there has been no indication that mitochondrial pH can be determined by NMR. The same is true of studies of perfused liver (McLaughlin et al. 1979; Iles et al. 1980). Only a single, relatively sharp inorganic phosphate ^{31}P resonance is observed, which is assigned to the cytoplasm. It seems unlikely that this inability to measure mitochondrial pH is due to inorganic phosphate being present at low concentration in mitochondria, because it is a substrate necessary for ATP synthesis in mitochondria. There are two possible explanations for this result. One is that the mitochondrial inorganic phosphate is in rapid exchange with cytoplasmic inorganic phosphate, or other phosphates (such as ATP), and so is not resolvable as a unique resonance. Another possible explanation is that the mitochondrial inorganic phosphate is relatively immobile, and so does not give a sharp, discrete signal; consistent with this view is the report that the concentration of inorganic phosphate measured in cell-free extracts of kidneys exceeds that observed in vivo (Freeman et al. 1983). Mention should be made of a report (Cohen et al. 1978), prefaced as preliminary, that separate ^{31}P-NMR lines due to cytosolic and mitochondrial inorganic phosphate can be seen in isolated liver cells. The validity of this report is not clear, since no further reports from the original authors, or other groups, have been forthcoming, in spite of the importance of the subject. It is possible that the two inorganic phosphate resonances observed in that study arose from subpopulations of cells differing in their physiological state, rather than pH differences within cells, since viabilities of 80–90% (initial) and 60–70% (final) were indicated by vital stain tests (Cohen et al. 1978). One might surmise, therefore, that the pH in compartments other than cytoplasm and vacuole may be difficult to measure by NMR even if a particular compartment occupies a sizeable fraction of the tissue volume.

11 Conclusions

A variety of NMR methods can be used to measure the pH in the cytoplasm and vacuoles in plant tissues. Unquestionably new approaches useful for determination of pH in specific plant species or tissues will be developed in the future, particularly using ^{13}C- and ^{15}N-labeled compounds. As with other methods for intracellular pH determination, there are a number of potential pitfalls that can limit the absolute accuracy of pH measurements. However, intracellular pH changes can be monitored with considerable precision and accuracy. Heterogeneity in pH over the sample can also be determined. The ability to make such measurements continuously on a single sample during a series of experimental conditions, and in some cases follow cytoplasmic and vacuolar pH simultaneously, are particular advantages of the method.

References

Bates RG (1981) The modern meaning of pH. CRC Crit Rev Anal Chem 10:247–278

Becker ED (1980) High resolution NMR: Theory and chemical applications. Academic Press, New York

Bishop SP (1980) Cardiovascular research. In: Baker HJ, Linsey JR, Weisbroth SH (eds) The laboratory rat, vol II. Academic Press, New York, pp 161–179

Brown FF, Campbell ID, Kuchel PW, Rabenstein DC (1977) Human erythrocyte metabolism studies by ^{1}H spin echo NMR. FEBS Lett 82:12–16

Cohen SM, Ogawa S, Rottenberg H, Glynn P, Yamane T, Brown TR, Shulman RG, Williamson JR (1978) ^{31}P nuclear magnetic resonance studies of isolated rat liver cells. Nature 273:554–556

Cramer JA, Prestegard JH (1977) NMR studies of pH-induced transport of carboxylic acids across phospholipid vesicle membranes. Biochem Biophys Res Commun 75:295–301

Deutsch C, Taylor JS, Wilson DF (1982) Regulation of intracellular pH by human peripheral blood lymphocytes as measured by ^{19}F-NMR. Proc Natl Acad Sci USA 79:7944–7948

Farrar TC, Becker ED (1971) Pulse and Fourier transform NMR: Introduction to theory and methods. Academic Press, New York

Foyer C, Walker D, Spencer C, Mann B (1982) Observations in the phosphate status and intracellular pH of intact cells, protoplasts and chloroplasts from photosynthetic tissue using phosphorus-31 nuclear magnetic resonance. Biochem J 202:429–434

Freeman D, Bartlett S, Radda G, Ross B (1983) Energetics of sodium transport in the kidney. Saturation transfer ^{31}P-NMR. Biochim Biophys Acta 762:325–336

Gadian DG (1982) Nuclear magnetic resonance and its application to living systems. Oxford University Press, Oxford

Hollis DP, Nunnally RL (1980) Recent ^{31}P-n.m.r. studies of myocardium. Phil Trans R Soc Lond Ser B 289:437–439

Hore PJ (1983) Solvent suppression in Fourier transform nuclear magnetic resonance. J Magn Reson 55:283–300

Iles RA, Griffiths JR, Stevens AN, Gadian DG, Porteus R (1980) Effects of fructose on the energy metabolism and acid-base status of the perfused starved-rat liver. Biochem J 192:191–202

Lee RB, Ratcliffe RG (1983) Phosphorus nutrition and the intracellular distribution of inorganic phosphate in pea root tips: A quantitative study using ^{31}P-NMR. J Exp Bot 34:1222–1224
Legerton TL, Kanamori K, Weiss RL, Roberts JD (1983) Measurements of cytoplasmic and vacuolar pH *Neurospora* using nitrogen-15 nuclear magnetic resonance spectroscopy. Biochemistry 22:899–903
Martin J-B, Bligny R, Rebeille F, Douce R, Lagauy J-J, Matheiu Y, Guern J (1982) A ^{31}P nuclear magnetic resonance study of intracellular pH of plant cells cultivated in liquid medium. Plant Physiol 70:1156–1161
Matile Ph (1978) Biochemistry and function of vacuoles. Annu Rev Plant Physiol 29:193–213
Matthews PM, Bland JL, Gadian DG, Radda GK (1981) The steady-state rate of ATP synthesis in the perfused rat heart measured by ^{31}P-NMR saturation transfer. Biochem Biophys Res Commun 103:1052–1059
McLaughlin AC, Takeda H, Chance B (1979) Rapid ATP assays in perfused mouse liver by ^{31}P-NMR. Proc Natl Acad Sci USA 76:5445–5449
Moon RB, Richards JH (1973) Determination of intracellular pH by ^{31}P magnetic resonance. J Biol Chem 248:7276–7278
Navon G, Ogawa S, Shulman RG, Yamane T (1977) High resolution ^{31}P nuclear magnetic resonance studies of metabolism in aerobic *Escherichia coli* cells. Proc Natl Acad Sci USA 74:888–891
Nuccitelli R, Webb DJ, Lagier ST, Matson GB (1981) ^{31}P-NMR reveals increased intracellular pH after fertilization in *Xenopus* eggs. Proc Natl Acad Sci USA 78:4421–4425
Ogino T, Arata Y, Fujiwara S (1980) Proton correlation nuclear magnetic resonance study of metabolic regulations and pyruvate transport in anaerobic *Escherichia coli* cells. Biochemistry 19:3684–3691
Pitman MG, Lauchli A, Stelzer R (1981) Ion distribution in roots of barley seedlings measured by electron probe X-ray microanalysis. Plant Physiol 68:673–679
Poole-Wilson PA (1978) Measurement of myocardial intracellular pH in pathological states. J Mol Cell Cardiol 10:511–526
Rebeille F, Bligny R, Martin J-B, Douce R (1983) Relationship between the cytoplasm and the vacuole phosphate pool in *Acer pseudoplatanus* cells. Arch Biochem Biophys 225:143–148
Roberts JKM (1984) Study of plant metabolism in vivo using NMR spectroscopy. Annu Rev Plant Physiol 35:375–386
Roberts JKM, Jardetzky O (1981) Monitoring of cellular metabolism by NMR. Biochim Biophys Acta 639:53–76
Roberts JKM, Ray PM, Wade-Jardetzky N, Jardetzky O (1980) Estimation of cytoplasmic and vacuolar pH in higher plant cells by ^{31}P-NMR. Nature 283:870–872
Roberts JKM, Ray PM, Wade-Jardetzky N, Jardetzky O (1981 a) Extent of intracellular pH changes during H^+ extrusion by maize root-tip cells. Planta 152:74–78
Roberts JKM, Wade-Jardetzky N, Jardetzky O (1981 b) Intracellular pH measurements by ^{31}P-nuclear magnetic resonance. Influence of factors other than pH in ^{31}P chemical shifts. Biochemistry 20:5389–5394
Roberts JKM, Wemmer D, Ray PM, Jardetzky O (1982) Regulation of cytoplasmic and vacuolar pH in maize root tips under different experimental conditions. Plant Physiol 69:1344–1347
Roberts JKM, Callis J, Wemmer D, Walbot V, Jardetzky O (1984 a) The mechanism of cytoplasmic pH regulation in hypoxic maize root tips and its role in survival under hypoxia. Proc Natl Acad Sci USA 81:3379–3383
Roberts JKM, Wemmer D, Jardetzky O (1984 b) Measurement of mitochondrial ATPase activity in maize root tips by saturation transfer ^{31}P nuclear magnetic resonance. Plant Physiol 74:632–639
Roberts JKM, Callis J, Jardetzky O, Walbot V, Freeling M (1984 c) Cytoplasmic acidosis as a determinant of flooding intolerance in plants. Proc Natl Acad Sci USA 81:6021–6033
Roos A, Boron WF (1981) Intracellular pH. Physiol Rev 61:296–434

Shaw D (1976) Fourier transform NMR spectroscopy. Elsevier, Amsterdam

Smith FA, Raven JA (1979) Intracellular pH and its regulation. Annu Rev Plant Physiol 30:289–311

Stidham MA, Moreland DE, Siedow JN (1983) ^{13}C nuclear magnetic resonance studies of crassulacean acid metabolism in intact leaves of *Kalanchoe tubiflora*. Plant Physiol 73:517–520

Ugurbil K, Brown TR, den Hollander JA, Glynn P, Shulman RG (1978) High-resolution ^{13}C nuclear magnetic resonances studies of glucose metabolism in *Escherichia coli*. Proc Natl Acad Sci USA 75:3742–3746

Ugurbil K, Shulman RG, Brown TR (1979) High resolution ^{31}P and ^{13}C nuclear magnetic resonance studies of *Escherichia coli* cells in vivo. In: Shulman RG (ed) Biological applications of magnetic resonance. Academic, New York, pp 537–589

Waterton JC, Bridges IG, Irving MP (1983) Intracellular compartmentation detected by ^{31}P-NMR in intact photosynthetic wheat-leaf tissue. Biochim Biophys Acta 763:315–320

Wray V, Schiel O, Berlin J (1983) High field phosphorus-31 nuclear magnetic resonance investigation of the phosphate metabolites in cell suspension cultures of *Nicotiana tabacum*. Z Pflanzenphysiol 112:215–220

Orientation of Chloroplasts in Leaves by ^{1}H NMR Spectroscopy

D. C. McCain

1 Introduction

1.1 What is NMR?

Certain atomic nuclei, including those of hydrogen atoms (^{1}H, or protons), possess energy levels that diverge in a magnetic field. The separation between these levels can be measured using the technique of Nuclear Magnetic Resonance (NMR) spectroscopy. An NMR spectrum is a plot of energy absorption versus resonance frequency, where frequency is proportional to the energy level separation. All ^{1}H levels diverge by the same amount in a uniform field. Therefore, when all nuclei in a sample experience the same magnetic field strength, their resonance frequencies are identical, and the NMR spectrum from that sample is a single "peak" at one specific frequency.

In a typical NMR experiment, a sample containing protons in a variety of chemical environments (e.g., at inequivalent sites in a chemical compound) is exposed to a constant, uniform magnetic field. Instead of a single peak, however, the resulting spectrum often displays a number of peaks at different resonance frequencies. The multiple peaks demonstrate that nuclei are exposed to different magnetic field strengths, even though a uniform field has been applied. Such a spectrum provides information about the magnetic environment at each nucleus; it can be used for chemical identification.

NMR also can be used to study larger structures. For example, zeugmatography (NMR imaging) has medical applications. In this technique, the sample (e.g., a human body) is composed primarily of water. The NMR spectrum of liquid water in a homogeneous magnetic field is a single peak. However, for zeugmatography, an inhomogeneous field is used, and the shape of the spectrum reveals the distribution of water throughout the sample as a function of field. Since field geometry is known, the spectral information can be used to reconstruct an image showing the positions of internal body organs.

This chapter describes an application of ^{1}H-NMR spectroscopy to studies of plant leaf structure. The method is somewhat similar to zeugmatography, because information about the physical structure of a leaf is encoded in the shape of the spectrum. However, a homogeneous applied field is used.

A leaf is mostly water; protons from other compounds in solution are too dilute to contribute significantly to the spectrum, and protons in the solid state or in membranes have peaks too wide to be detected. Therefore, the ^{1}H-NMR spectrum of a leaf is essentially that of water, and in a homogeneous magnetic field, we may expect the spectrum to be a single peak. Most leaves do show a single peak, however, the spectrum from some leaves presents a complex, multi-peak pattern showing that leaf water is present in different magnetic environments. We must conclude that some leaves are magnetically nonuniform, i.e., the field is distorted by structures inside the leaf.

Although the NMR spectrum of a leaf contains structural information, the information is conveyed in a format different from that of an image. The spectrum must be interpreted by means of model calculations, and the information can be recovered only as numerical data. Also, unlike an image, the information represents a statistical average of data obtained simultaneously from all parts of the sample.

1.2 NMR Spectra of Plant Leaves

^{1}H-NMR spectroscopy has seldom been used for direct studies of plant tissue; perhaps its most important use has been for studies of freezing injury (Burke et al. 1976). Examples of NMR research on leaf tissue include studies of freezing injury in wheat leaves (Macdowall and Buchanan 1974) and of manganese translocation in maize leaves (Bacic and Ratkovic 1984). In studies of this kind, the typical spectrum is a broad, single peak, and nuclear spin relaxation measurements often provide evidence showing that water is present in more than one compartment.

Complex (i.e., multiple-peak) NMR spectra have been observed in plant tissue, but usually the results reflect magnetic field distortions caused by the shape of the sample. For example, Burke et al. (1976) obtained a doublet spectrum from dogwood stems and assigned the peaks to water protons in the bark. Their calculations show that the field inside the sample is distorted to produce a doublet spectrum when the sample has the shape of a hollow cylinder. A dogwood stem is like a hollow cylinder because the bark contains water while the inner stem is dry. Lenk et al. (1980, 1981) report a doublet NMR spectrum from spinach leaves; they assigned the doublet to nuclear dipole-dipole interactions in ordered water. However, their results also could be ascribed to effects of sample geometry because the leaves were rolled into a cylinder before being inserted into the NMR sample tube.

McCain et al. (1984) obtained ^{1}H-NMR spectra from the leaves of 50 plant species. They found that the leaves of most species produce a single NMR peak, as expected, but that leaves from about one-fourth of the species show complex spectra in which two or three peaks can be discerned. Great care was taken to control sample-shape effects; the sample holder was designed to provide a homogeneous magnetic field. Figures in this chapter present some previously unpublished spectra obtained using this technique.

Figure 1 shows that the spectrum of a spinach leaf (*Spinacea oleracea*) is a single, symmetrical peak. Other plant species with similar spectra include: alfalfa (*Medicago sativa*), and pea (*Pisum sativum*). The same spectrum is obtained from these species regardless of leaf orientation with respect of the applied magnetic field. The peak width (about 2 ppm at half height) is much greater than one would expect from a pure water sample. Probably a large part of the excess width is

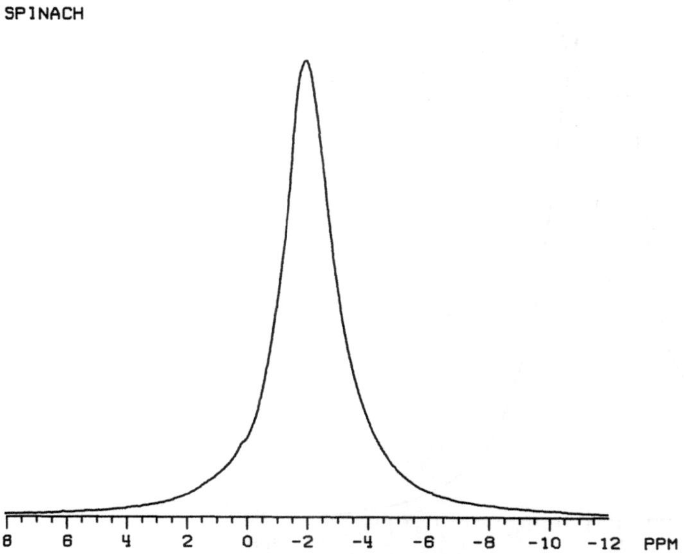

Fig. 1. The ¹H-NMR spectrum of a spinach leaf (*Spinacea oleracea*) obtained at room temperature with a spectrometer frequency of 470 MHz. A disc 4-mm in diameter was cut from a fresh spinach leaf and placed in a special sample holder designed to insure that the leaf surface was perpendicular to a homogeneous, applied magnetic field. The origin of the chemical shift scale (at 0 ppm) is arbitrary because no chemical shift reference was present in the sample, but 0 ppm corresponds approximately to the position of the bulk water signal

caused by the presence of air spaces between water-filled compartments inside the leaf. The magnetic susceptibility of air is different from that of water, and the nonuniform medium disturbs the homogeneity of the applied field. This broadens the spectrum. Separate experiments show that imperfections in the experimental apparatus contribute less than 0.4 ppm to the total peak width, and saturation experiments confirm that much of the peak width is caused by inhomogeneous broadening. Figure 1 tells us very little about the internal structure of a spinach leaf except that it is heterogeneous.

Figure 2 presents the spectrum of a fresh hawthorn leaf (*Crataegus* sp.) obtained with the leaf surface oriented perpendicular to the applied magnetic field. The spectrum was recorded only a few minutes after the leaf had been picked from the tree. Clearly, there are at least three peaks. To facilitate discussion, these have been labeled A, B, and C, beginning on the left-hand (low field) side.

Reasonably well-resolved ABC patterns have been found in leaves from a number of species (approximately one-fourth of the total surveyed), including: pawpaw (*Asimina triloba*), redbud (*Cercis canadensis*), and persimmon (*Diospyros virginiana*). Each species has its own characteristic pattern distinguished from those of other species by different relative peak intensities, different spacings between the peaks, and different peak widths. In some species the pattern is highly reproducible, in others it is more variable. About the only common feature in all patterns is that peak A is always lower in amplitude than peaks B and C. The

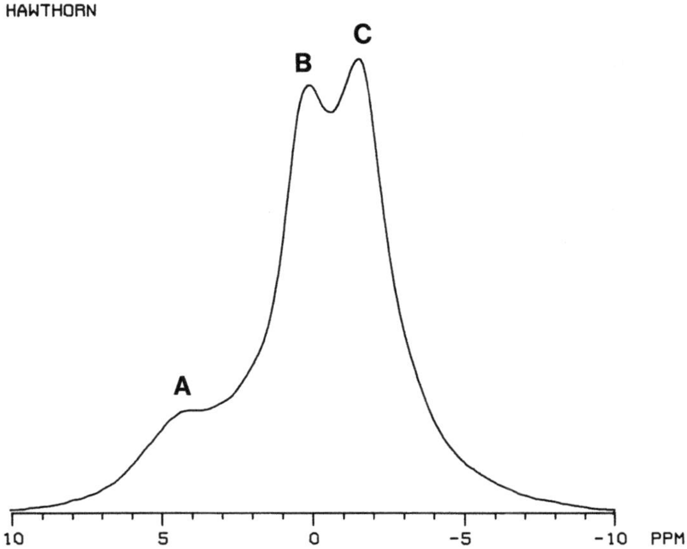

Fig. 2. The ^{1}H-NMR spectrum of a hawthorn leaf (*Crataegus* sp.) obtained under exactly the same conditions as Fig. 1. Three peaks (*A*, *B*, *C*) can be discerned

ABC pattern may be present but poorly resolved in a majority of plant species; about half the species examined have single peaks with asymmetric shapes that suggest an unresolved ABC pattern.

Among the species examined, hawthorn provides the best example of an ABC pattern, because the peaks are unusually well resolved and reproducible. The relative intensities and spacings are quite consistent from one hawthorn sample to another. Also, peaks A and C are unusually strong in hawthorn relative to peak B; they are typically much smaller in other species.

To summarize the results, most species showing well-resolved ABC patterns are trees, and the best results seem to come from leaves grown in full sunlight. Plants with single, symmetrical peaks are typically herbs with thick leaves and a high water content (e.g., spinach and pea). Species with asymmetric peaks are not as easily characterized; they include all grasses, many trees (including most species of oak) and many thin-leafed herbs.

It is possible to demonstrate that different peaks in the ABC pattern are signals from different water compartments. Experiments with a number of species show that peak B decreases in amplitude relative to peaks A and C as a leaf dries. One explanation for this effect is that the peaks represent signals from water in different leaf compartments, and that different compartments lose water at different rates as the leaf dries.

Relaxation measurements provide additional evidence to show that the ABC peaks are produced by water in different compartments. The nuclear spin-lattice relaxation times (T_1) for water protons are sensitive to the local environment. Water molecules diffuse throughout a small compartment in times comparable to T_1, so all water protons in a given compartment may be expected to have the same relaxation time. However, measured T_1 values are different for different peaks; this means that the water molecules are isolated from each other (McCain et al. 1984). Saturation-transfer experiments have been interpreted to show that the protons responsible for each NMR peak are isolated on a short time-scale from the others, but on a longer time-scale, proton exchange occurs between compartments (McCain and Markley 1985 a).

The ABC pattern varies as the angle between the leaf surface and the applied magnetic field is varied. Figure 3 displays a series of experimental spectra obtained from a hawthorn sample at orientations from $\theta = 0°$ to $90°$. θ is defined as the angle between the applied magnetic field and a vector normal to the leaf surface. Examining Fig. 3, we see that the normal ABC pattern (at $\theta = 0°$) changes shape dramatically as the leaf is rotated in the field. By 30°, peaks A and C converge on the position of peak B; peak C barely can be seen as an unresolved

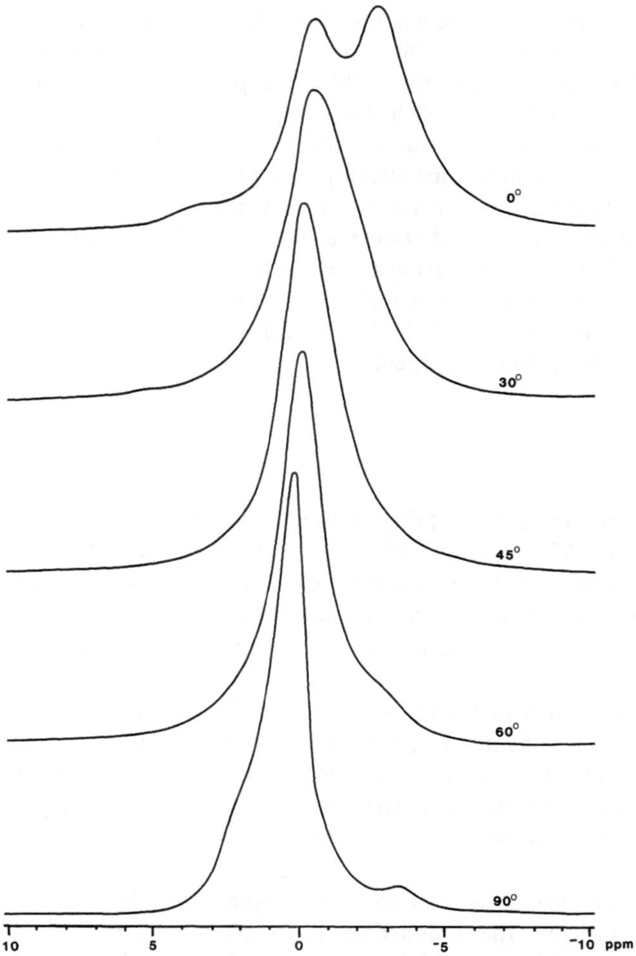

Fig. 3. A series of 470 MHz NMR spectra obtained from a single hawthorn leaf disc at various orientations with respect to the applied magnetic field. The angle, θ, is defined as that between the magnetic field and a vector normal to the leaf surface

shoulder. At 45 °, the three peaks are superimposed. Partial resolution returns at 90 °; at this orientation, a weak, high-field peak appears near the 0 ° position of peak C, and a stronger low-field shoulder may be seen near the 0 ° position of peak A. Relaxation time measurements demonstrate that peak C (the high-field peak at 0 °) crosses through the spectrum to become the low-field shoulder at 90 °.

2 Experimental Technique

2.1 Choosing an NMR Spectrometer

A high-field NMR spectrometer is best for leaf studies, even though the experimental sensitivity and resolution requirements are modest. The high field is needed for several reasons: (1) Field-independent relaxation mechanisms (e.g., interactions with transition-metal ions and nuclear dipole-dipole splittings from ordered water) degrade the resolution at low field when peak separations are small; (2) chloroplast water protons produce a single, resolvable peak only when the proton spin-spin relaxation time, T_2, is substantially shorter than the lifetime for proton exchange through the chloroplast envelope membrane (ca. 80 ms) but longer than the time required for proton diffusion across a chloroplast (ca. 3 ms). Since T_2 contains a field-dependent term that decreases as the field increases, the range of suitable fields is limited. Considering these factors, the optimum magnetic field for many leaf samples approximately corresponds to the range of superconducting NMR spectrometers (200 to 500 MHz). In many cases, however, lower fields may provide satisfactory spectra.

2.2 Sample Holders

A special sample holder, one that does not distort the field, is required for these experiments. Conventional NMR sample tubes alone cannot be used. If a leaf were rolled or folded and inserted into an ordinary NMR sample tube, the magnetic field inside the sample could be distorted by sample-shape effects. The resulting NMR spectrum would be complex, even if the sample were internally homogeneous.

To eliminate shape effects, it is necessary to use a sample that has the shape of an ellipsoid. In practice, it is convenient to use an ellipsoid with one or two infinite dimensions, (i.e., a plane or a cylinder) that has been truncated to the shape of a disc or a rod. To a good approximation, the apparatus described below provides a uniform applied field because the disc-shaped leaf sample is completely encased in a glass rod.

Since NMR spectrometers are designed to accept sample tubes of a standard size, a sample holder that fits inside the standard tube is most convenient. Suitable sample tube inserts have been constructred from 4-mm, OD Pyrex glass rod selected for a close fit to the inner diameter of a standard NMR sample tube. The rod length is not critical, but lengths of ca. 16 cm, approximately the length of the sample tube, are suitable. The rods should be rounded (or beveled) at one end to conform approximately to the hemispherical inner surface at the closed end of an NMR sample tube. The rods may then be cut through at a point that corresponds to the center of the probe coil when the sample tube and insert are inside the spectrometer. A cut point 1.5 cm from the beveled end is about right, but the best position may vary from one spectrometer to another, due to differences in probe design. Cut surfaces should be polished smooth.

Each insert rod orients a disc-shaped leaf sample at one specific angle to the applied magnetic field. A set of rods cut at precise angles of 90 °, 60 °, 45 °, and 30 ° to the rod axis is convenient. Since in superconducting magnets the field runs parallel to the sample tube axis, this set of inserts orients the leaf sample so that a vector normal to its surface makes an angle (defined as the angle θ) of 0 °, 30 °, 45 °, or 60 °, respectively, to the magnetic field. A slightly different insert assembly is needed to study samples aligned with leaf surfaces parallel to the sample tube axis (i.e., at $\theta = 90$ °). A suitable holder was constructed from a 4-mm length of rod that had been split down its axis; a leaf sample can be sandwiched between the two split halves and held between the flat ends of rods cut at an angle of 90 ° to the axis. The sample holders described above are suitable only for superconducting magnets where the external field is parallel to the sample tube axis; they are not designed for an iron-core magnet that has its magnetic field perpendicular to the sample tube axis.

Sample holders should be tested to demonstrate that any complex NMR spectra that may be observed are not experimental artifacts. One may use a disc-shaped control sample made from any suitable material that may be expected to display a narrow, single-peak ^{1}H-NMR spectrum. At every angle, the control should give a single peak with constant width and no evidence of unresolved structure. Peak widths of ca. 1 ppm have been obtained from a disc of moist filter paper; the peak is broadened by the magnetically heterogeneous medium (fibers, air, and water), but no evidence for multiple peaks can be found. A disc cut from a mushroom gill provided an even better control sample; the single peak was considerably narrower, only 0.4 ppm wide. Therefore, the broadening caused by the sample holder must be less than 0.4 ppm.

As the control sample is rotated in the magnetic field, using different sample tube inserts, the peak center may shift progressively; for example, rotation through a 90 ° angle shifts the peak from filter paper by about 1 ppm. This happens when the disc and the insert rod have different magnetic susceptibilities; in that case the magnetic field inside the disc is homogeneous, but it is offset by an amount that depends on orientation.

2.3 Sample Preparation

Leaf discs are cut from fresh leaves using a sharp 4-mm cork borer (available from chemical supply companies). To cut a disc, one first selects a region on the leaf avoiding large veins. The leaf is placed against a soft backing, and the disc is cut out with a quick twisting motion of the cork borer. It is important that the sample disc be quite dry on its surface and at its cut edges. Moisture on the sample (extracellular water) will produce an additional signal in the NMR spectrum – generally a much sharper signal than that obtained from intracellular leaf water. Moisture may be blotted from the leaf surface if necessary, but leaves that "bleed" at the cut edges (e.g., spinach) make poor samples. A small bump due to extracellular water may be seen in Fig. 1 at about 0 ppm on the arbitrary chemical shift scale.

It is possible to obtain an NMR spectrum less than 3 min after the leaf sample has been harvested from the plant. However, such haste is probably unnecessary. No apparent changes have been seen in the spectrum while the leaf disc remains in the sample tube for periods of up to 1 h; the sample cannot dry out because it is completely encased in glass. Leaves can be field-collected, kept moist, and stored in a humid refrigerator for as long as 1 day with good results. Leaves attached to branches with the cut ends immersed in water may be kept for several days in a lighted room, but spectra from individual leaves or leaf discs exposed to warm, dry air change rapidly as the samples dry.

2.4 Operating the Spectrometer

Operating procedures for leaf samples follow standard NMR practice with some minor differences. In the following discussion, any procedures not specifically mentioned may be assumed to be similar to those used with conventional samples. Adjust the spectrometer as follows:

1. Adjust the magnet shims in the usual way to obtain a homogeneous field, using a conventional reference sample in a standard NMR sample tube. Do not use the leaf sample when shimming because it is too small in size to serve as an accurate measure of field homogeniety. The magnet need not be shimmed to extremely high resolution because leaf samples give broad peaks.
2. Leaf samples do not produce a lock signal, therefore one must turn off the internal field-lock circuit. The lock is not needed for these experiments because a persistent superconducting magnet will not drift significantly in the short time required to obtain a leaf spectrum. If the spectrometer has an external lock system, it may be used.
3. The sample spinner may be turned off if desired. The peaks from leaves are inherently broad and cannot be narrowed by spinning.
4. The best pulse sequence is a series of single pulses (with pulse angles from 60 ° to 90 °) separated by delay times much longer than T_1, the spin-lattice relaxation time.
5. Choose a recycle time (acquisition time plus pulse delay) of four or more T_1 periods. Typical T_1 values for leaf water protons are in the range from 0.1 to 1 s; if T_1 has not been measured, a delay time of from 2 to 5 s is a reasonable choice.
6. Using the leaf sample, determine and set the proper pulse width (pulse angle). Do not assume that it is the same value as required for conventional samples because the sample-tube insert is likely to change the probe response.
7. A spectral window with a width of about 20 ppm is suitable. Since resolution is generally low, it is not necessary to acquire many data points from the free induction decay (FID); 1 K or 2 K data points are adequate.

8. Use a small number of pulses to acquire the spectrum. Typically, the signal from a leaf is so strong that excellent signal-to-noise rations may be obtained in one minute or less.

9. Before performing the Fourier transform, some manipulation of the FID may be helpful. Exponential apodization (filtering) may be used to remove high-frequency noise on the spectrum; this is not really necessary because the signal-to-noise ratio is high, but it produces a cleaner spectrum.

10. A problem often encountered in leaf spectra is a very broad background signal that lifts the water spectrum above the baseline. When this broad signal is present, peak intensities are difficult to measure accurately. There are several ways to eliminate the broad peak and discriminate in favor of the relatively narrow signals. The easiest way is simply to shift the baseline up to the bottom of the narrow peaks. Another method is to increase the delay interval between the transmitter pulse and the acquisition of the FID. The same effect can also be produced by applying a "left shift" to the data in the FID; a left shift discards the first data point (obtained immediately after the transmitter pulse) and shifts all other data points down in time. Yet another way to emphasize the narrow peaks is to use a trapezoidal or sinusoidal apodization; this method distorts peak shapes and relative intensities, but it can be useful for such purposes as measuring peak splittings.

NMR is a versatile technique, and many different experiments are possible. For example, one can measure saturation-transfer, T_1, or T_2. Detailed instructions can be found in appropriate textbooks (Becker 1980) and need not be repeated here since a leaf sample would be treated very much like any other. Note, however, that such experiments require fairly well-resolved spectra.

3 Data Analysis

3.1 The Need for a Theory

Clearly, the NMR spectrum of a leaf contains information about its structure. Some of the information can be recovered using NMR techniques that require no specific knowledge about leaves. For example, the following statements are supported by experimental data: (1) Relaxation, saturation-transfer and drying experiments all show that different spectral peaks are signals from protons in different water compartments. (2) The relative positions of peaks A and C invert about peak B as the leaf sample is rotated in the magnetic field. This suggests that peaks A and C originate from water in magnetically anisotropic compartments that are oriented with respect to the leaf surface. (3) Peak splittings vary with temperature and with the concentration of manganous ions in the leaves. Both these effects suggest that paramagnetic ions cause the shifts that separate the peaks. (4) NMR spectral patterns are more or less reproducible within a single plant species, but not from one species to another. This fact suggests that we shall find an explanation for the patterns in some feature of leaf structure or leaf chemistry that varies dramatically with species.

The conclusions outlined above tell us something about the state of water in a leaf, but nothing we can relate directly to leaf structure. To extract more information, it is necessary to analyze the spectra using a theory developed specifically for leaves. A successful theory would explain the origin of the orientation-dependent ABC pattern in terms of leaf anatomy and physiology. It would provide answers to such questions as: What causes the splittings that separate the peaks? Or, why does each species have its own characteristic spectral pattern? Finally, to be useful, the theory should relate the NMR results to subjects of general interest to botanists and plant physiologists.

3.2 Review of Leaf Anatomy and Physiology

Before describing the theory, we shall review briefly some aspects of leaf structure that may effect the NMR spectrum (Esau 1960).

A dicotyledon leaf is composed of parallel layers of cells; these are the epidermal, palisade, and spongy mesophil layers. Epidermal and stomatal cells occupy the leaf surface. The palisade layer, immediately below the upper surface, consists of narrow, cylindrical cells with axes directed perpendicular to the leaf surface. Cells in the spongy layer are irregularly shaped. Both the palisade and spongy layers contain intracellular air spaces, with the largest fraction of air space in the spongy layer.

Leaf water is contained in relatively isolated compartments separated by membranes. Each cell is a compartment, and within it there are subcompartments; the chloroplasts, vacuole, and cytoplasm. Extracellular water also may be present. Chloroplasts contain a significant fraction of the total leaf water, often more than 25% of it.

Chloroplasts are irregular in shape, but approximately they correspond to ellipsoids with three unequal dimensions. As such they may be described in terms of three perpendicular axes, with two of the axes defined by the longest and the shortest distances through the center of the chloroplast. The thylakoid membranes inside a chloroplast are all aligned parallel to each other and perpendicular to the shortest axis. These features describe an organelle that is intrinsically anisotropic.

Most chloroplasts are located in the palisade layer, although some are found in the stomatal and spongy mesophyl cells. Typically, the chloroplasts adhere to or are pressed against the cell walls in such a way that the shortest chloroplast axis is perpendicular to the wall. A large fraction of wall area in the palisade cells is perpendicular to the leaf surface; therefore the chloroplasts in a leaf are aligned nonrandomly.

Chloroplasts accumulate manganese in several distinct chemical environments or "pools" (Amesz 1983). One pool is at the active sites for water splitting in photosystem II; this pool has roughly constant concentration in all plants (about 1 Mn atom per 100 chlorophyll molecules). Another pool is the "reserve" or "weakly bound" manganese; these are Mn^{2+} ions bound to specific sites on the outer surface of the thylakoid. The function of reserve manganese has not been established definitely; it may act as a superoxide dismutase. In any case, the quantity of Mn^{2+} in the reserve pool is extremely variable. Some species (e.g., spinach) have very little, while other species have much higher concentrations (Theg and Sayre 1979). The few species tested that are known to have large amounts of reserve manganese show complex NMR spectra, while those known to have small amounts give single peaks.

3.3 The Theory

A theory has been devised to explain the NMR spectrum of a leaf (McCain and Markley 1985b). The theory attempts to calculate the average magnetic field to which water molecules are exposed in the different cell compartments. The assumptions below are intended to define and simplify the problem; to be useful, they need not be strictly true as long as they are approximately correct. Starting assumptions are as follows:

1. The NMR spectrum of a leaf is the spectrum of its water protons. Signals from other protons in the leaf may be ignored; either the other protons are present in too low a concentration, or their peaks are too wide to be detected.
2. All water protons have the same NMR chemical shift and would produce a single NMR peak if they were in a uniform magnetic field. However, when water molecules in different regions of a leaf experience different local magnetic field strengths, the water signal will be split into several peaks.
3. Each water compartment in a leaf contributes a single peak to the NMR spectrum. Within a small compartment, diffusion thoroughly mixes all water protons during the time interval T_2, the proton spin-spin relaxation time. T_2 represents the averaging time over which a nucleus responds to its local environment. Mixing insures that all the water protons are magnetically equivalent (on an NMR time scale); therefore the spectrum of an individual chloroplast (or other small water compartment) is a single peak at a position that represents its average internal magnetic field.
4. The average magnetic field experienced by a water proton inside a chloroplast depends on the concentration and distribution of manganous ions. The local magnetic field is the sum of three terms: the external applied field, the time-averaged sum of dipolar fields from reserve manganous ions embedded in the thylakoid membrane and a susceptibility correction that depends in part on the local concentration of manganous ions. The dipolar field is anisotropic; its value depends on the orientation of the thylakoid membrane with respect to the external magnetic field. It is possible to estimate the size of these terms and to conclude that the spectral positions of chloroplast water protons could be shifted by as much as the largest experimental peak splittings when reserve Mn^{2+} concentrations are high. However, it is impossible at this time to fully interpret the splittings because the detailed structure of the thylakoid membrane is unknown.
5. The spectral pattern obtained from one plant species appears different from that of another because of differences in internal leaf structure. The most important factors are: the relative magnetic susceptibilities of adjacent cell layers, the concentrations of reserve manganese in the chloroplasts, the distribution of chloroplast orientations with respect to the leaf surface, and the fraction of total leaf water present in the chloroplasts.

3.4 The Model

The model is a description of an idealized leaf structure that is suitable for mathematical analysis. It has been simplified until it contains the minimum number of adjustable parameters necessary to reproduce the general features of experimental leaf spectra; therefore it is useful, but certainly it is not a complete or accurate description of the internal structure of real leaves. The model is defined in the following statements (McCain and Markley 1985 b):

1. The model leaf is composed of two cell layers, the palisade and spongy mesophil; the epidermal layer has been ignored because it contains only a small fraction of the total leaf water. Palisade cells in the model leaf have the shape of long cylinders with axes perpendicular to the leaf surface. Chloroplasts are attached to the walls of the palisade cells. We shall define f_c as the fraction of the total leaf water present inside chloroplasts that are pressed against the vertical walls of the palisade cells (i.e., with the shortest chloroplast axes parallel to the leaf surface). Another set of chloroplasts containing water fraction f_a is pressed against the ends of the palisade cells (i.e., with short axes perpendicular to the leaf surface). In a typical leaf we may expect f_c to be greater than f_a because the palisade cells are longer than they are wide.

2. The average magnetic field experienced by water protons inside a chloroplast is assumed to be the sum of the applied field (corrected for susceptibility) and an anisotropic tensor function. For convenience, values of the tensor components may be expressed as ppm shifts from the spectral position of bulk water protons. We shall define the component parallel to the shortest chloroplast axis as α while the component parallel to the longest axis is defined as γ. The intermediate component is β. Since the tensor is assumed to be purely anisotropic, the sum of all three components is exactly zero. α, β, and γ cannot be calculated directly, since we do not know the detailed molecular structure of a thylakoid membrane; therefore we shall treat them as experimental variables.

3. The magnetic field distribution inside a layered structure depends on magnetic susceptibility differences between the layers and on the orientation of the layers with respect to an external field. We shall define the difference between the average magnetic susceptibility of the spongy mesophil and that of the palisade layer as δ. Accordingly, we may expect the signal from water in the palisade layer to be split from that of water in the spongy mesophil by an amount proportional to δ. Also, we shall define f_b as the fraction of total leaf water in the spongy mesophil.

The model outlined above is sufficient to produce orientation-dependent ABC patterns similar to those observed experimentally. By varying the adjustable parameters, it is possible to generate a theoretical spectrum that approximately reproduces the experimental spectrum from any given leaf sample. This can be accomplished in spite of the fact that many important structural features in real leaves have been omitted from the model.

For example, a real leaf contains more water fractions than the three defined in the model (f_a, f_b, and f_c). Presumably, the model succeeds because the other water fractions are of minor importance in the NMR spectrum of a real leaf. Some of the water fractions in a real leaf are negligible in comparison to the total water content (e.g., the stomatal cells); others may not appear as separate peaks because their signals overlap those of the three model fractions (e.g., epidermal cells and the cytoplasm and vacuoles of palisade cells).

The model also assumes that chloroplasts have only two possible orientations, but obviously real leaves are more complex. Signals from chloroplasts at intermediate orientations would appear at intermediate positions in the spectrum (between peaks A and C). Perhaps this explains why many experimental spectra are poorly resolved. They may be produced by leaves containing mostly disoriented chloroplasts. Such spectra would be broad and assymmetric but otherwise unresolved (e.g., experimental spectra from grasses). Other unresolved spectra may be caused by chloroplasts that have small anisotropic tensor components due to low concentrations of reserve manganese (e.g., spinach); these would appear as a single, nearly symmetrical peak.

3.5 Calculations

Using principles of physics, it is possible to calculate the spectrum of water molecules occupying compartments in the model described above. The spectrum is given by a rather complicated equation (McCain and Markley 1985b). Rather than reproduce the equation here, a computer program is listed that performs the calculations. The program, reproduced in Fig. 4, is written in the BASIC language. The program requires seven input parameters:

1. The anisotropic tensor components, α and γ (in ppm units). These are the extreme values of the field shift obtained when the chloroplasts are aligned so that their short and long axes, respectively, are aligned perpendicular to the applied magnetic field. α Normally is a negative number, while γ is positive. The magnitudes of α and γ depend on the concentration of reserve manganese in the chloroplasts, and in some poorly understood way on the structure of the thylakoid membrane. It is not possible to predict the values of α and γ for a particular leaf; they must be determined by experiment.

2. The orientation-dependent field offset, δ (in ppm), is a measure of the difference in magnetic susceptibility between the palisade and spongy mesophil layers. δ may have either positive or negative values, but for theoretical reasons, we expect it to range between -0.7 ppm and the value of γ.

3. The water fractions f_a and f_c (in percent units) are the fractions of total leaf water that occur in chloroplasts oriented with short axes perpendicular and parallel, respectively, to the leaf surface. f_a and f_c are proportional to the heights of peaks A and C, respectively.

4. The peak width at half height, W, in ppm units.

5. The orientation of the leaf with respect to the magnetic field, θ (in degrees).

To calculate a leaf spectrum at any given angle, θ, the model requires six input parameters: α, γ, δ, f_a, f_c, and W. Five parameters or combinations of parameters can be measured directly from the experimental spectrum at $\theta = 0$ °. The peak width can be measured directly if the resolution is sufficient to separate individual peaks; but even with poor resolution, one can estimate the peak width. The mea-

```
1 REM  ** THIS PROGRAM COMPUTES THE NMR SPECTRUM OF A LEAF **
2 REM BASED ON A MODEL WITH ORIENTED, ANISOTROPIC CHLOROPLASTS.
10 DIM R(73),S(173)
20 PRINT "ANISOTROPIC TENSOR COMPONENTS (PPM): ALPHA, GAMMA =";
30 INPUT A,G
40 PRINT "ORIENTATION-DEPENDENT FIELD OFFSET (PPM): DELTA =";
50 INPUT D
60 PRINT "CHLOROPLAST WATER FRACTIONS (PERCENT):PEAK A,PEAK C";
70 INPUT F1,F2
80 PRINT "PEAK WIDTH AT HALF HEIGHT (PPM) =";
90 INPUT L
100 PRINT "ANGLE BETWEEN LEAF AND MAGNETIC FIELD (DEGREES) =";
110 INPUT T
120 C=COS(T*.017453)**2
130 FOR I=.01 TO 1.57 STEP .02
140 P=24*(1+G/A)*COS(I)**2-24*SIN(I)**2
150 Q=24*(1+G/A)*COS(I)**2-24*G*SIN(I)**2/A
160 R(INT(P*(1-C)-24*G*C/A+25))=R(INT(P*(1-C)-24*G*C/A+25))+F2
170 R(INT(Q*(1-C)-24*C+25))=R(INT(Q*(1-C)-24*C+25))+F1
180 NEXT I
190 R(INT(25-24*D*C/A))=R(INT(25-24*D*C/A))+79*(100-F1-F2)
200 FOR I=1 TO 73
210 S(I+50)=S(I+50)+R(I)
220 FOR J=1 TO 122
230 IF (I+J)>123 GOTO 250
240 S(I+J+50)=S(I+J+50)+R(I)*144/(144+(J*A/L)**2)
250 IF (I-J)<-49 GOTO 270
260 S(I-J+50)=S(I-J+50)+R(I)*144/(144+(J*A/L)**2)
270 NEXT J
280 R(I)=0
290 NEXT I
300 PRINT "OUTPUT OPTIONS: PLOT ONLY=0, PRINT ONLY=1, BOTH=2";
310 INPUT K
320 IF K=1 GOTO 370
330 FOR I=26 TO 149 STEP 3
340 PRINT TAB(INT(S(I)/110))"*"
350 NEXT I
360 IF K=0 GOTO 450
370 PRINT "FIELD STRENGTH (PPM*1000) AND SPECTRAL INTENSITY"
380 PRINT "FIELD";TAB(8);"INTENSITY";TAB(26);"FIELD";TAB(34);
390 PRINT "INTENSITY";TAB(50);"FIELD";TAB(58);"INTENSITY"
400 FOR I=1 TO 57
410 PRINT INT(41.67*A*(I-75));TAB(10);INT(S(I));TAB(26);
420 PRINT INT(41.67*A*(I-18));TAB(34);INT(S(I+57));TAB(50);
430 PRINT INT(41.67*A*(I+39));TAB(59);INT(S(I+114))
440 NEXT I
450 FOR I=1 TO 173
460 S(I)=0
470 NEXT I
480 PRINT "RECYCLE?: NEW ANGLE=1,NEW WIDTH=2,ALL NEW=3,STOP=4";
490 INPUT K
500 ON K GOTO 100,80,20,999
999 END
```

Fig. 4. A computer program that can be used to simulate the theoretical NMR spectrum of a model leaf. Given appropriate input data, the program produces either a plot of the spectrum or a table of spectral data points. Spectra obtained from this program are shown as *dotted lines* in Fig. 5, 6, and 7

sured heights of peaks A, B, and C are proportional to f_a, f_b, and f_c, respectively; numerical values (as percent of total leaf water) may be determined by setting the sum of f_a, f_b, and f_c to 100%. At $\theta = 0°$ the spacing between peaks A and C is equal to the difference $\gamma - \alpha$ while the spacing from peak B to peak C is equal to $\gamma - \delta$.

A strategy for matching experimental NMR spectra to calculations might involve the following steps: (1) Using a spectrum obtained at $\theta = 0°$, measure the peak heights and spacings. From these derive first estimates for the values of f_a, f_c, $\gamma - \alpha$, and $\gamma - \delta$. Estimate W. (2) Assume a value of γ. Use this to generate a calculated spectrum. If the fit is to be made only to a spectrum at $\theta = 0\,°$, it is not necessary to determine the value of γ; otherwise, vary γ while keeping the other parameters constant until the match is fairly good. Finally, (3) refine the fit by varying, slightly, the values of all parameters to find the optimum combination. Since the model is greatly simplified, one must not expect the final fit to be perfect.

After the program has been loaded into the computer, one enters the appropriate data. For example, to calculate the spectrum of *Helianthus tuberosus* (Fig. 5), the following parameters were used: $\alpha = -2.4$ ppm, $\gamma = 3.4$ ppm, $\delta = 3.4$ ppm, peak A = 12%, peak C = 24%, and width = 2.1 ppm. After the data have been entered, the computer will ask for output options. It can either plot the data graphically for direct comparison with the NMR spectrum or it can print a set of data points (peak heights vs. field).

If a satisfactory fit can be obtained, we are assured that the model is consistent with the data, although a good fit does not prove the model is correct. In addition, the fit provides experimental parameters that describe the leaf spectrum. The significance of some parameters is still obscure; α, β, γ, and δ are related in some way to the structure of cell layers, the topology of the thylakoid membrane, and the

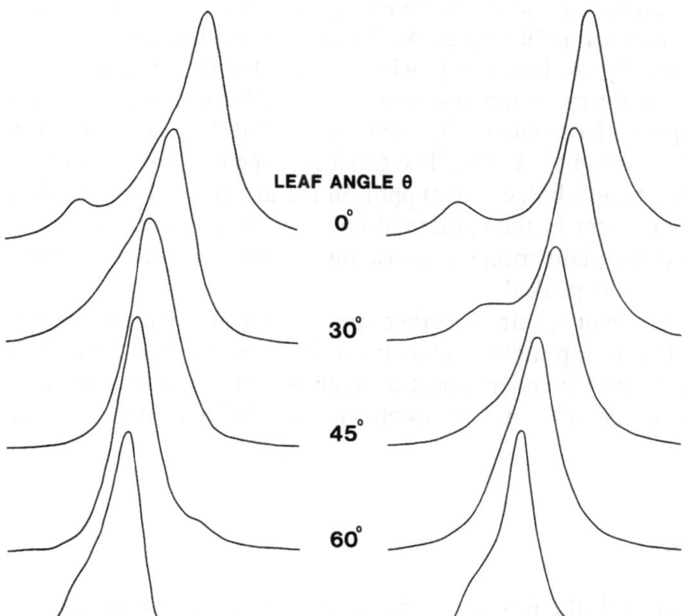

Fig. 5. A comparison of experimental orientation-dependent NMR spectra with model calculations. Experimental spectra (*left*) were obtained from *Helianthus tuberosus* leaves at five different orientations in the magnetic field. Model calculations (*right*) were performed using input parameters described in the text. The figure shows that model calculations reproduce most of the orientation-dependent features observed in experimental spectra

concentration of reserve Mn^{2+} ions. At this time it is not clear how to use this information.

Probably the most valuable information in the spectrum is in the relative peak heights. Peak heights are easily measured; a computer is not needed when the spectrum is well-resolved. According to the model, the heights of peaks A and C reveal the fraction of the total leaf water in chloroplasts that have specific orientations with respect to the leaf surface, while peak B represents the water in all other leaf compartments. Therefore, peak heights can be used to study the distribution of water in a leaf as well as to study the statistics of orientation in a large population of chloroplasts. Such information would be difficult to obtain without the use of NMR.

4 Examples of Interpreted Spectra

4.1 Orientation Dependence

Figure 5 compares experimental leaf spectra (left side) from Jerusalem artichoke (*Helianthus tuberosus*) with results of model calculations (right side). The fit is not perfect, but all major features in the experimental spectra are reproduced by the model. *H. tuberosus* is unusual in having peaks B and C superimposed at $\theta = 0°$. The three peaks are partially resolved at 90 ° where, according to the model, peak B appears in the center of the spectrum and peaks A and C have crossed through the center to sides opposite from where they were located at 0 ° (i.e., at 90 ° the order of peaks from left to right is CBA). The general shape of the 0 ° spectrum is well reproduced except for a bulge near 0 ppm on the arbitrary chemical shift scale. The bulge may be water in the epidermal layer; a similar bulge is seldom noted in spectra from other plant species because the central portion of the spectrum usually is dominated by peak B.

The orientation-dependent spectra of other species also fit the model rather well, but with quite different input parameters from those used in Fig. 5. General success in fitting orientation-dependent spectra suggests that the most important parameters necessary to describe a leaf spectrum have been included in the model.

4.2 Species Dependence

Figure 6 illustrates how well the model can be made to fit experimental spectra from a number of plant species. Note that the shape of each spectrum is quite different and that the model parameters (Table 1) for each species also are quite different, while the parameters from different samples of the same species (e.g., different samples of *Crataegus*) are rather similar. The spectrum of *L. tulipifera* has only two peaks; peak A is always very weak or absent in this species. The spectra shown are representative of their species; resolution does not vary dramatically from one specimen to another.

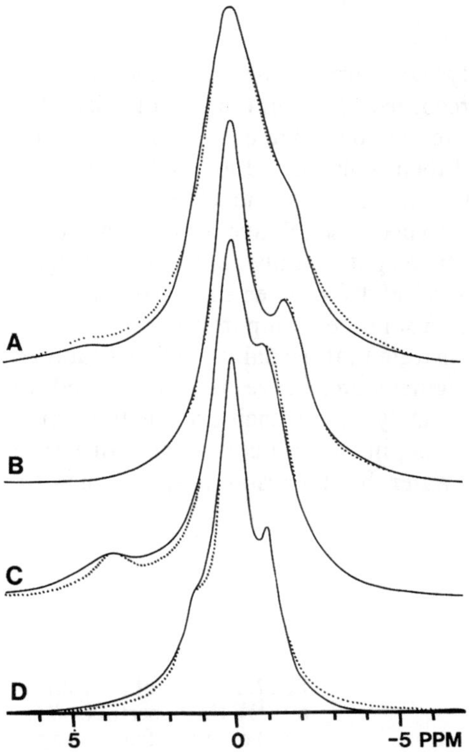

Fig. 6. A comparison of experimental species-dependent NMR spectra with model calculations. Experimental spectra (*solid lines*) at $\theta = 0°$ were obtained from four different plant species: *A Quercus bicolor* (swamp oak); *B Liriodendron tulipifera* (tulip poplar); *C Acer negundo* (boxelder); and *D Asimina triloba* (pawpaw). Computer calculations (*dotted lines*) were performed using the model parameters listed in Table 1. Calculated spectra represent the best fit to experimental spectra within the constraints of the model. This figure illustrates the large differences found between spectra from different plant species

Table 1. Model parameters obtained from computer simulations of experimental leaf spectra[a]

Species	Figure no.	W	f_a	f_c	$\gamma-\alpha$	$\gamma-\delta$
Spinacea oleracea	1	2.0	0	0	–	–
Crataegus sp.	2	2.3	11	47	6.0	1.8
Crataegus sp.	3	2.2	7	45	6.2	2.4
Crataegus sp.	8 (A)	2.7	12	40	6.2	2.0
Helianthus tuberosus	5	2.1	12	24	5.8	0.0
Quercus bicolor	6 (A)	2.1	3	21	6.3	1.7
Liriodendron tulipifera	6 (B)	1.3	0	29	–	1.7
Acer negundo	6 (C)	1.4	6	33	5.0	1.3
Asimina triloba	6 (D)	1.0	14	23	2.3	1.1
Aesculus glabra (fresh)	7 (A)	1.5	22	39	2.6	1.5
A. glabra (dried 10 min)	7 (B)	1.4	22	42	2.7	1.5
A. glabra (dried 30 min)	7 (C)	1.5	22	50	3.0	1.5
A. glabra (dried 50 min)	7 (D)	1.8	24	58	3.0	1.4

[a] The model parameters represent optimum values to fit the experimental spectra shown in Figs. 1–8, using the computer program listed in Fig. 4. Some of the computed spectra are shown as dotted lines in the figures, others are not

4.3 Effects of Drying

Figure 7 illustrates a use of NMR to study water stress. Trace (a) is the spectrum of a fresh buckeye leaf (*Aesculus glabra*) recorded a few minutes after the leaf was harvested. Trace (b) is the same sample after it had been removed from the NMR sample tube, dried for 10 min in air, and then replaced in the NMR. Note that peak B has decreased in intensity relative to peaks A and C. Traces (c) and (d) continue the trend. Peak B continues to decrease relative to peaks A and C. All four traces were scaled to the same total height, so only relative peak heights can be measured. We may conclude that as a leaf dries, water is lost from the chloroplasts (peaks A and C) less rapidly than from other compartments.

Other experiments (not shown) demonstrate that partially dried leaf samples can be revived to approximately their original state. Leaves that have dried for a short while (up to about trace c) approximately recover their original fresh spectrum (trace a) after a short period of soaking in water. Leaves dried for longer periods (trace d and beyond) also absorb water, but the resulting spectrum is different from that of the fresh leaf.

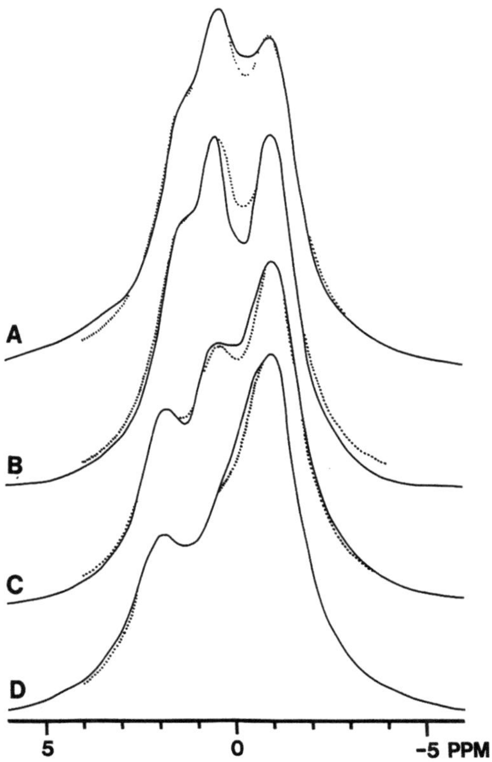

Fig. 7. A series of experimental ^{1}H-NMR spectra (*solid lines*) obtained from a single leaf disc of *Aesculus glabra* (buckeye) at $\theta = 0$ °: *A* the spectrum of the fresh sample, taken immediately after the leaf was harvested; *B* after the leaf disc had dried in air for 10 min; *C* after drying a total of 30 min; *D* after drying for 50 min. For comparison, a set of calculated spectra (*dotted lines*) is shown; these were obtained using the parameters listed in Table 1. Note that as the leaf dries, the central peak decreases in amplitude relative to the peaks on either side. This trend is reflected in the parameters of Table 1 which show that water fractions f_a and f_c increase as the leaf dries

4.4 Effects of Freezing

Figure 8 illustrates the effects of freezing on leaf spectra. In contrast to Fig. 7, all traces were plotted on the same intensity scale, so peak areas are proportional to the number of water protons contributing to the signal. Only the signal from fluid water protons can be seen. The spectrum of ice is extremely broad and cannot be seen on this scale.

Figure 8(a) is the spectrum of a hawthorn leaf at a temperature of 5 °C. It is indistinguishable from a spectrum of the same sample obtained at 20 °C. While in the NMR probe, the sample was subjected to a stream of cold nitrogen at −11 °C, and spectra were recorded as the sample froze. Trace (c) is one of the intermediate spectra obtained during freezing. Its intensity is lower than (a) because much of the water has turned to ice. The shape of (c) suggests that water in peak B freezes before that in peaks A or C. There are several possible explanations for the result. It may be that the freezing point of water in the chloroplasts is lower than that for water in one or more of the other compartments, or (more

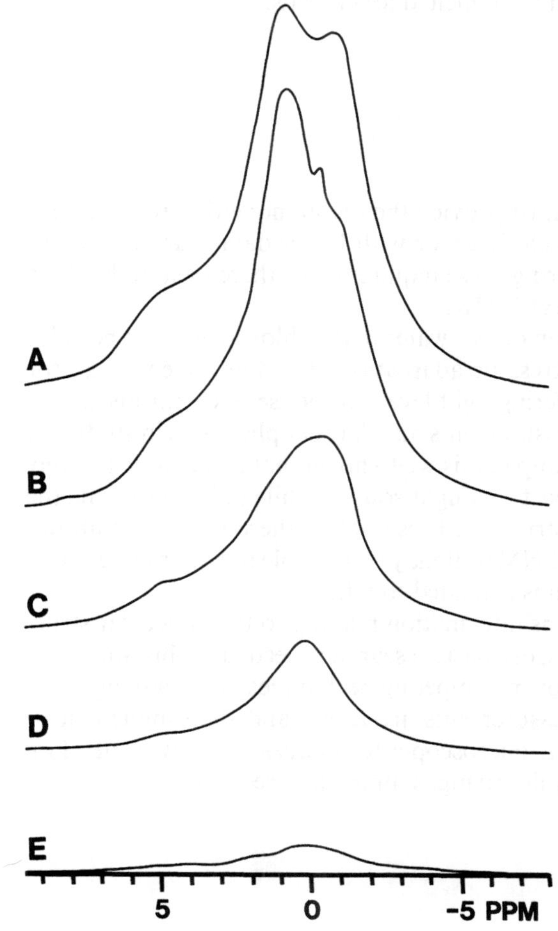

Fig. 8. A cycle of ^{1}H-NMR spectra obtained from a single hawthorn leaf disc (*Crataegus* sp.) as it froze and then thawed: *A* the spectrum of the fresh leaf (before freezing) obtained at a temperature of 5 °C; *B* the spectrum of the thawed leaf (after warming from −11 °C) obtained at 5 °C; *C* a spectrum obtained during the freezing process as the leaf cooled from 5 °C to −11 °C; *D* obtained during the thawing process as the leaf warmed from −11 °C to 5 °C; *E* the spectrum of the frozen leaf at −11 °C. A comparison of trace (*A*) with trace (*B*) shows that the leaf has sustained damage as a result of freezing

likely) the effect is caused by differences in freezing rate and chloroplast water does not freeze as fast as does water in the other compartments (perhaps it lacks crystallization nuclei).

Figure 8 (e) is the spectrum of the frozen hawthorn leaf at −11 °C. It shows that most of the water is ice, but that a small fraction of the water remains mobile. The same observation has been made in many plant tissues (Burke et al. 1976). At low temperatures, mobile water is presumably associated with macromolecules that change its structure and prevent freezing.

The frozen sample, while still in the probe, was subjected to a stream of nitrogen gas at 5 °C. Trace (d) was obtained during the melting process. It seems to show that water melts at the same rate in all compartments. Trace (b) is the final spectrum of melted leaves at 5 °C. It shows that the intensity of peak C is lower than the corresponding peak in the spectrum of a fresh leaf. This probably indicates that some of the chloroplast envelope membranes have broken so that some of the chloroplast water no longer is isolated from that of the surroundings. Trace (b) also shows a small, sharp peak that has the same appearance as would be produced by a drop of extracellular water. Probably it is the signal from small droplets of pure water that formed from melted ice crystals.

5 Future Directions

In order to be useful, leaf NMR must provide the researcher either with new information inaccessible by other techniques or with better data than can be obtained by other methods. Considering these requirements, there seem to be three most promising applications for leaf NMR.

1. Measurements of the fraction of leaf water in the chloroplasts can be valuable in studies involving water stress or adaptations to environment (e.g., for studies of leaf development in differing light levels, response to chemicals, etc.).

2. Chloroplast orientation measurements may have applications in studies of photosynthetic efficiency. For example, it is well known that algae often actively orient their chloroplasts with respect to a light source, while in higher plants the orientation is built into the leaf structure. Presumably, there must exist an optimum orientation. With the aid of NMR, it may be possible to better understand how order in photosynthetic systems is related to efficiency.

3. The NMR spectrum contains information relating to the molecular structure of the thylakoid membrane. More basic research is needed in this area.

NMR has some advantages over competing techniques for studying chloroplasts. For example, NMR measurements are rapid, and the sample can be healthy, living tissue. By contrast, microscopic techniques are slow (one chloroplast at a time), and usually require fixing, staining and sectioning.

References

Amesz J (1983) The role of manganese in photosynthetic oxygen evolution. Biochim Biophys Acta 726:1–12
Bacic G, Ratkovic S (1984) Water exchange in plant tissue studied by proton NMR in the presence of paramagnetic centers. Biophys J 45:767–776
Becker ED (1980) High-resolution NMR. Academic Press, New York
Burke MJ, Gusta LV, Quamme HA, Weiser CJ, Li PH (1976) Freezing and injury in plants. Annu Rev Plant Physiol 27:507–528
Esau K (1960) Anatomy of seed plants. Wiley, New York
Lenk R, Bonzon M, Greppin H (1980) Dynamically oriented biological water as studied by NMR. Chem Phys Lett 76:175–177
Lenk R, Bonzon M, Greppin H (1981) Irreversible thermodynamics and biological evolution in spinach leaves as studied by NMR. Z Pflanzenphysiol 101:107–118
Macdowall FDH, Buchanan GW (1974) Estimation of the water of hydration in wintering wheat leaves by proton magnetic resonance. Can J Biochem 52:652–654
McCain DC, Markley JL (1985a) Water permeability of chloroplast envelope membranes: in vivo measurement by saturation transfer NMR. FEBS Lett 183:353–357
McCain DC, Markley JL (1985b) A theory and a model for interpreting the proton NMR spectra of water in plant leaves. Biophys J 48:687–694
McCain DC, Selig TC, Govindjee, Markley JL (1984) Some plant leaves have orientation-dependent EPR and NMR spectra. Proc Natl Acad Sci USA 81:748–752
Theg SM, Sayre RT (1979) Characterization of chloroplast manganese by electron paramagnetic resonance spectroscopy. Plant Sci Lett 16:319–326

^{13}C-NMR Determination of Rubber Content in Guayule Bushes

J. Visintainer and R. C. Hirst

1 Introduction

1.1 Background

The commercial cultivation of the guayule bush as a source of natural rubber (cis-1,4-polyisoprene) is seriously being considered (Nat Acad Sci Report 1977). In order to obtain the rubber, the complete bush is harvested and the rubber is extracted with a solvent. There is a need for analytical methods to determine the rubber content at various stages of the processing; i.e., after chopping, after milling, after removal of the resin, and after extraction of the rubber. In addition, guayule bushes will be cultivated under a variety of conditions to try to improve the rubber content. For such work it may be desirable to have an analytical method which can determine the rubber content in a small portion of one branch or of one root.

Hevea rubber and guayule rubber have the same chemical structure. That is, both are high cis-1,4-polyisoprenes with nondetectable amounts of other structural units (Nat Acad Sci Report 1977). In this report the term "rubber" is used to refer to such high cis-1,4-polyisoprenes.

It is desirable to rapidly measure the rubber content in guayule bushes for the following reasons: (a) to select seeds from bushes that yield higher rubber contents; (b) to follow the rate of rubber production in the bush; and (c) to measure residual rubber content after extraction in any pilot plant or commercial extraction process.

1.2 Analytical Methods

Several methods are currently available for the determination of the rubber content of guayule bushes. Solvent extraction methods typically require about 10 h and 20 g of sample, and extraction results represent an average for a whole sample. Infrared (IR) spectrometry methods have recently been reported (Gutierrez and Ransaw 1980; Banigan et al. 1981) which require that the bush be finely ground in order to obtain a uniform sample. These methods were reported as being applicable for samples with rubber contents in the 5–25% range. Other methods include a gravimetric method (Gowans and Clark 1952) and a turbidimetric method (Hanson and Perry 1981).

A brief carbon-13 (^{13}C) nuclear magnetic resonance (NMR) study, which estimated the rubber contents in milled guayule shrubs with rubber contents in the 2.5–5 wt% range, was carried out by Shoolery (personal communication), but was not published. A limitation of his approach was the requirement that a known weight of natural rubber be milled uniformly into each sample to provide an internal standard. The ^{13}C-NMR method reported here is an extension of our original work (Visintainer et al. 1981). This method does not require an internal standard in the sample, is not limited to a particular range of rubber contents, takes only 20 min of instrument time, and can be used on a branch segment clipped from a living bush. An alternate approach using an internal standard was taken by Hayman et al. (1982).

2 Experimental

2.1 ^{13}C-NMR Spectra

The ^{13}C-NMR spectra were obtained at ambient temperature on a Varian CFT-20 NMR spectrometer using the 10-mm ^{13}C probe and wide-band 1H decoupling. The spectral width was chosen to be 4000 Hz with an acquisition time of 0.2 s. The optimum flip angle of 90 (pulse width = 18 μs) was used along with a pulse delay of zero seconds. A minimum of 6000 scans (20 min) was co-added. The resulting signal was computer smoothed with a negative weighting factor of 0.05, which eliminated much of the spectral noise, yet did not broaden the spectral lines.

The relatively low-field CFT-20 was chosen because it is a commonly available spectrometer, easily accessible to a large number of potential users. A larger-field spectrometer would shorten the time required to perform the analysis, and would increase the precision, but the cost of purchasing such an instrument would also drastically increase.

Spin-lattice relaxation time (T_1) measurements were obtained using an 180-τ-90 pulse sequence (Breitmaier and Voelter 1974) where τ varied from 0.01 s to 2 s, and the pulse delay after the 90 pulse was 2.5 s (which included a 1-s acquisition time). The nuclear Overhauser effect (Breitmaier and Voelter 1974) behaved normally for polyisoprene systems (Schaefer 1972) and thus was not suppressed. All spectra were plotted by using the same internally consistent plot parameters for each spectrum. Electronic integration was used to measure all spectral areas in this work.

2.2 Sample Preparation

2.2.1 Guayule Material

Portions of branches, stems, roots, and unmilled, chopped bushes were examined as received. The maximum size of sample was 8 mm in diameter and 8 mm in length.

All ground or milled samples were first dried at 90 °C for ½ hour, thoroughly mixed, weighed, and then made into pellets. A die for compressing guayule

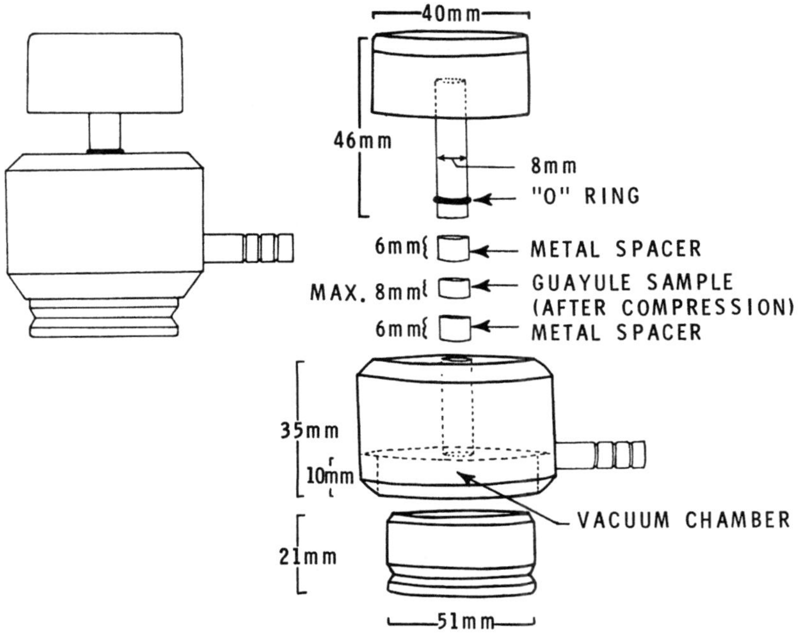

Fig. 1. Die used for compressing guayule samples to the volume specifications of the ^{13}C-NMR method

samples to our exact volume specifications (a cylinder approximately 8 mm long and 8 mm in diameter) was needed. A schematic of the die is shown in Fig. 1. Approximately 0.3 g of sample was added to the die, and then compressed for several minutes.

2.2.2 Sample Geometry

The rf coil geometry is shown in Fig. 2a. The guayule sample geometry, in a 10 mm ^{13}C-NMR sample tube, is shown in Fig. 2b. The ^{13}C tube was positioned such that the center of the guayule pellet was at the center of the rf coil. Vortex plugs were placed both above and below the guayule pellet. D_2O was added as a lock. The vortex plugs were stripped of their vanes to allow a greater volume of D_2O into the tube.

2.3 Calibration Curve

Calibration standards were made from a 1-mm-thick cured, non-carbon-black loaded natural rubber sheet that contained 93.5% by weight rubber. Eight samples of various weights were prepared. Each sample was placed in a "filler" of ground guayule material, from which the rubber had been extracted, and centered in the rf coil. This assured the same "Q" of the coil (Farrar and Becker

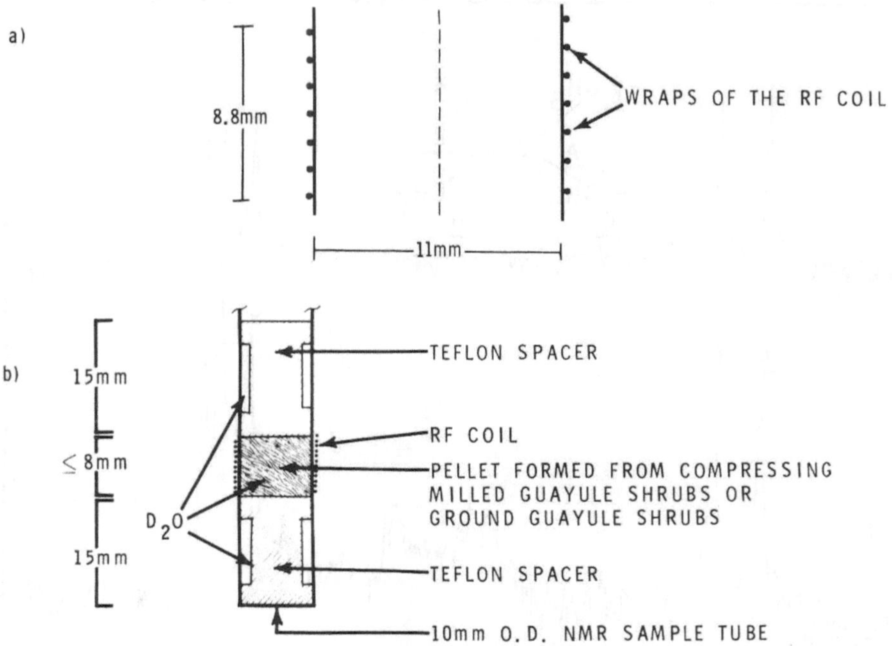

Fig. 2. a Geometry of the ^{13}C-NMR rf coil. **b** 10-mm ^{13}C-NMR tube and sample geometry. The tube is positioned so that the center of the guayule pellet is at the center of the rf coil. D_2O surrounds the spacers and pellet

1971). The NMR spectra were obtained according to the same instrument conditions used for the guayule measurements.

3 Results and Discussion

The measurements of the rubber contents in guayule bushes presented many formidable problems for ^{1}H-NMR spectroscopy due to the nonhomogeneities resulting from the presence of various solids in the bush. These solids include the soluble resins and the insoluble biopolymers (Shoolery, personal communication) along with the rubber, present in its amorphous phase. These difficulties were greatly decreased by measuring the ^{13}C-NMR spectrum of the bush.

3.1 ^{13}C-NMR Spectra

Spectra for cured natural rubber and for a typical milled, compressed sample of a guayule bush are shown in Fig. 3. The five cis-1,4-polyisoprene peaks, designated A, B, C, D, and E (Fig. 3), and the spin-lattice relaxation times (T_1), are listed in Table 1. The T_1's for the cured sample were 0.05 s for B, C, and D; and

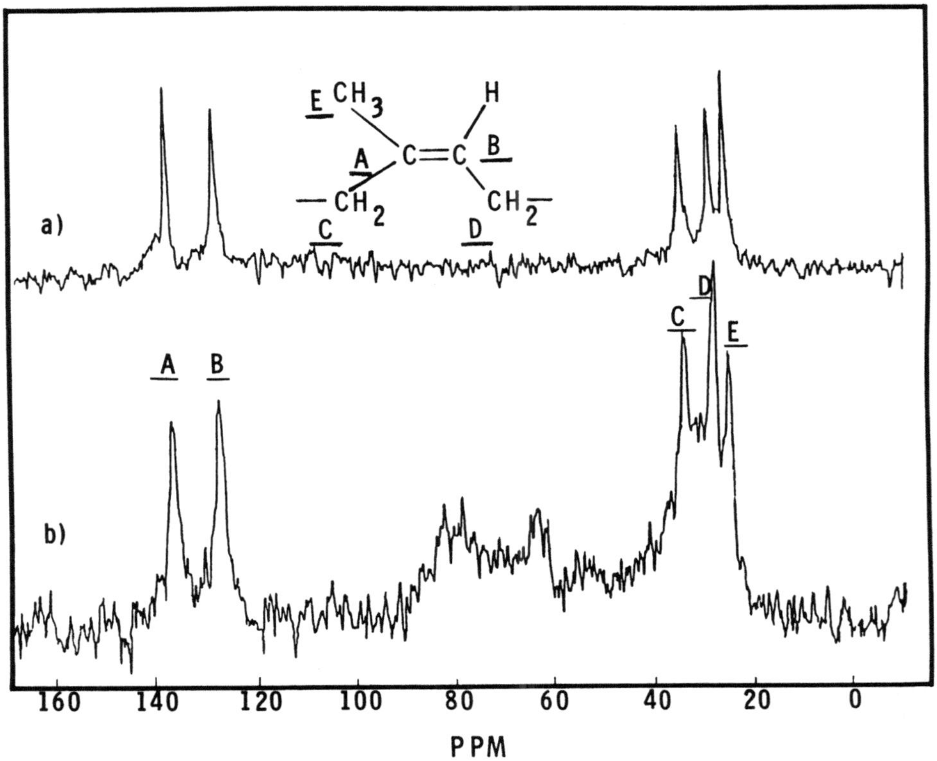

Fig. 3 a, b. 20 MHz spectrum, with full quantitative conditions, of **a** cured rubber, **b** guayule sample

0.4 to 0.5 s for A and E. The rubber peaks in the guayule samples gave similar T_1 values.

The background peaks due to the nonrubber materials in the guayule bush were studied. After a chopped bush was milled, its resin was removed through an acetone extraction. The ^{13}C spectra of this resin, and the same resin with a small piece of added cured rubber, are shown in Figs. 4 and 5, respectively. The material from which this resin was removed was then extracted with cyclohexane ten times

Table 1. Assignments and spin-lattice relaxation times (T_1) for cured sample of cis-1.4-polyisoprene

Peak	ppm from ME_4S_i	T_1, s
A	135	0.4
B	126	0.05
C	33	0.05
D	27	0.05
E	24	0.5

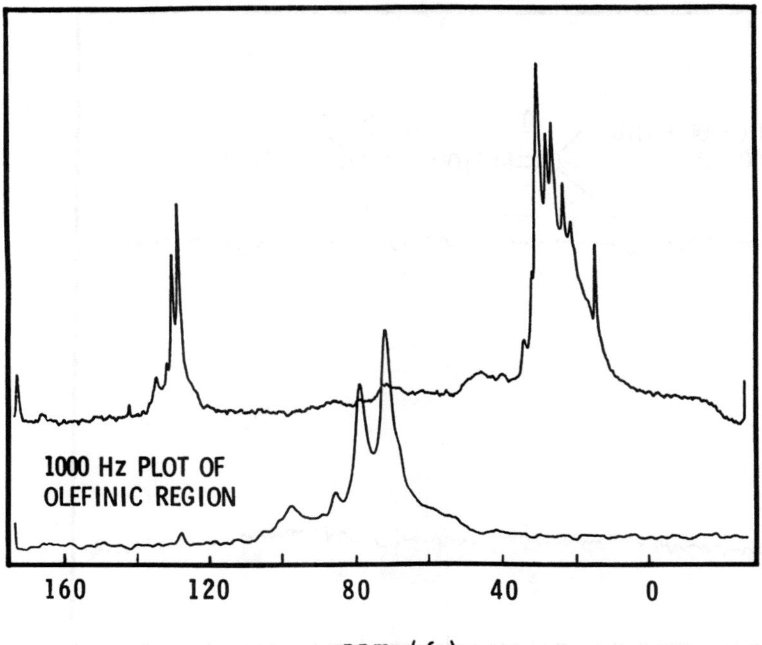

Fig. 4. 20 MHz ^{13}C-NMR spectrum of guayule resin, extracted with acetone, under quantitative conditions. Expansion of the olefinic region is also shown

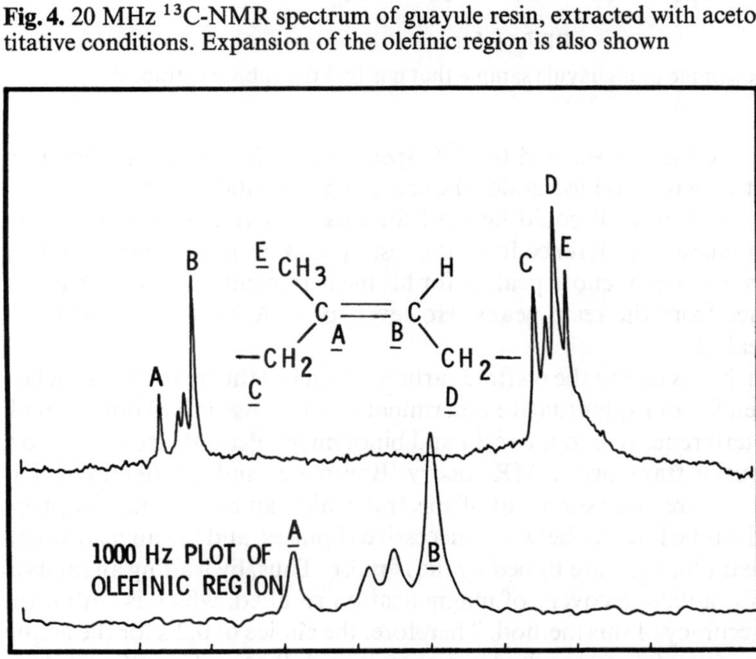

Fig. 5. 20 MHz ^{13}C-NMR spectrum of the acetone-extracted resin (see Fig. 4) with a rubber sample added. Expansion of the olefinic region is also shown

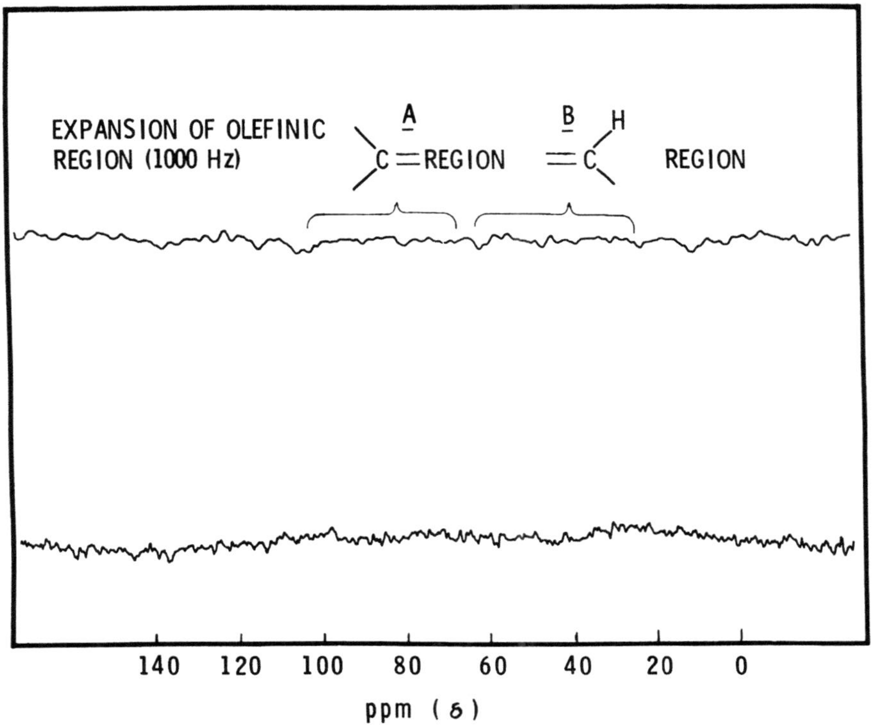

Fig. 6. ^{13}C-NMR sample of a guayule sample that has had the rubber extracted

in order to remove the rubber, and the ^{13}C spectrum of this sample is shown in Fig. 6. All spectra were obtained under the same experimental conditions.

Either peak A or peak B could be used for quantitative determinations because there is minimal interference from the resin peaks in those areas. Shoolery (personal communication) chose peak A for his measurements because it had the least interference from the resin peaks. However, peak A had a T_1 eight times greater than peak B.

Peak B, which was due to the olefinic carbon which has the hydrogen attached to it, was chosen for our quantitative determinations because it had only a small background interference due to the resin and bipolymers. Peak B also has a short T_1 (0.05 s). Fourier transform NMR theory (Breitmaier and Voelter 1974) has shown that correct area measurement of spectral peaks can be obtained by properly choosing the time interval between successive rf pulses, and the angle through which the nuclear moments are tipped by the rf pulse. Thus, by waiting an interval of four times T_1, a 98% recovery of magnetization resulted, which is within the experimental accuracy of this method. Therefore, the choice of 0.2 s for the acquisition time (four times 0.05 s) resulted in a correct peak B area. This effect is demonstrated in Fig. 7. Note that peak B gives full spectral area, while peak A, with its longer T_1, is "saturated" relative to peak B. The resin also had a longer T_1 in the olefinic region (about 1 s), and thus the repetition time of 0.2 s saturated its peak area as well, and gave a negligible contribution to the peak B area.

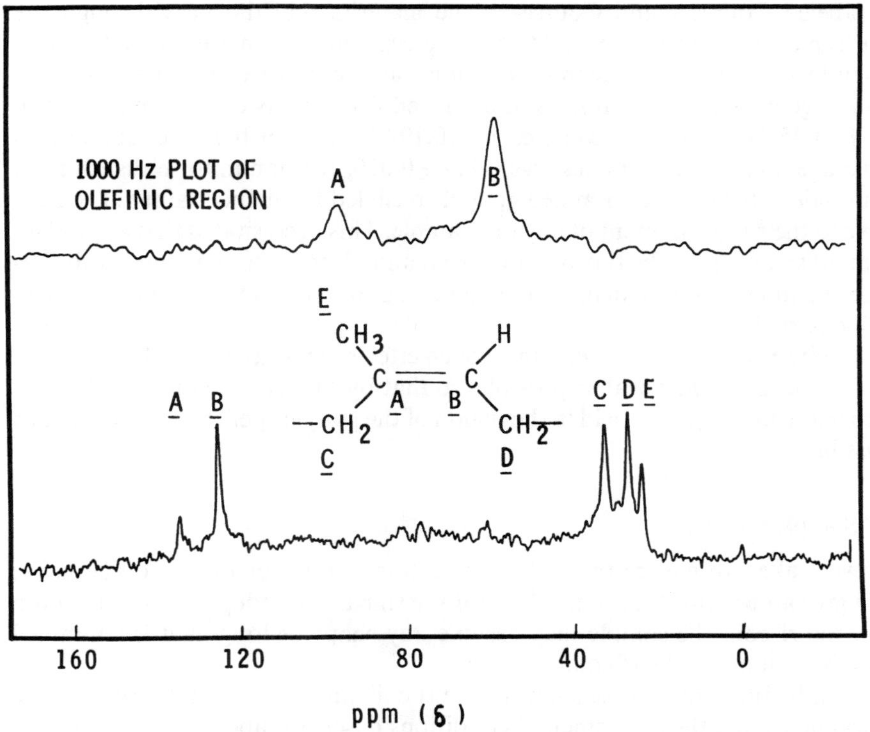

Fig. 7. 20 MHz ¹³C-NMR spectrum of a guayule sample, under the experimental conditions described in this report. Area *B* is quantitatively correct.

3.2 Sample Preparation

3.2.1 Guayule Sample

The weight percent rubber in a guayule sample could be determined through two NMR procedures: (1) The lengths of the guayule samples are the same size or smaller than that of the volume contained within the NMR rf coil. In this way, the amount of any rubber in any sample can be directly compared to the weight of the sample. This assumes that the NMR signal does not vary as a function of position within the rf coil. (2) The length of the sample is longer than the rf coil in order to obtain a better signal-to-noise ratio because of the higher filling factor.

Procedure (2) leads into several experimental difficulties. One was the necessity of having a uniform packing density in order to accurately relate the rubber content detected by the rf coil to the total weight of the sample. Experimentally the density of a ground or milled guayule sample was nonuniform, i.e., the rubber, resin, and biopolymers were not necessarily uniformly distributed in the sample. This was mainly due to the different settling gradients of large and small pieces.

As an illustration of the above point, two samples were made from a partially extracted bush whose material did not pass through a 10-mesh wire screen, and were labeled Course-1 and Course-2. The material that did pass through the coarse screen was placed on a 14-mesh screen, and the material that passed through this screen was made into two more samples labeled Fine-1 and Fine-2. The average areas for the samples, normalized for weights of the samples, were 0.232 ± 0.005 for the Coarse samples, and 0.192 ± 0.005 for the Fine samples. As the average area for all samples was 0.212 ± 0.020, the nonuniform distribution of materials in the ground or milled guayule bush led to an uncertainty of $\pm 10\%$ relative to the rubber content of any one sample. This necessitated that any milled or ground sample must be thoroughly mixed and placed into the sample tube before any settling occurred, in order to assure a uniform distribution of both coarse and fine particles.

Procedure (1) was used to eliminate such effects. This geometry (Fig. 2b) was chosen to be as symmetrical as possible, so that by tuning up on the D_2O one is able to make the magnetic field in the region of the guayule pellet as homogeneous as possible.

3.2.2 Sample Geometry

Care was taken to ensure that all cross-sectional areas of the rf coil (Fig. 2a) would give an equal NMR signal. Thus if the signal was independent of position within the rf coil, the sample may assume any shape as long as it is contained within the volume of the rf coil.

To study different cross-sections of the rf coil, an 8-mm diameter rubber disc was scanned using the experimental conditions described above, with the exception that the repetition time was 0.4. The disc weighed 0.059 g and was cut from a 1-mm thick cured rubber sheet containing no carbon black. The disc was placed in the rf coil with its surface perpendicular to the axis of the coil, and its peak B area was plotted as a function of position within the rf coil. The results are shown

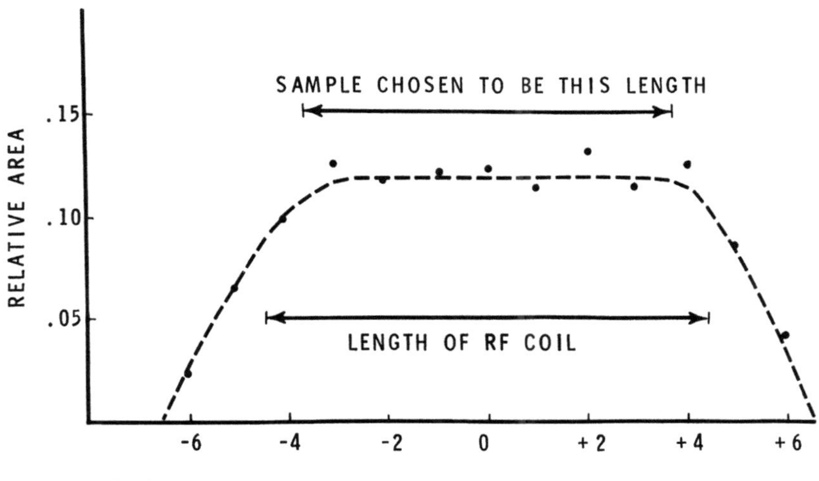

Fig. 8. Area of the cis-1,4-polyisoprene peak due to the methine olefinic carbon vs. the position of a 1-mm thick gum rubber disc (0.059 g) in the rf coil

in Fig. 8. Thus, if the guayule sample is kept to a maximum length of 8 mm and centered within the rf coil (as shown in Fig. 2), the signal from any section of the sample within the coil will be quantitatively and directly related to the signal from all sections of the sample.

A minor problem with the above geometry was that the guayule pellet tended to swell by about 20% of its volume after contact with the D_2O. Since the D_2O is needed as a magnetic field lock, we must take into account the D_2O swelling of the sample as the pellet is being made. The ideal would be to use an *external* lock, eliminating the need for D_2O, in the sample tube.

3.2.2.1 Calibration Plot

The plot of peak B versus weight of rubber is given in Fig. 9 for various standard rubber discs centered in the rf coil (Table 2). The plot was a straight line, and within experimental uncertainty the intercept passed through the origin.

Care must be taken when comparing the relative NMR signals from a rubber disc centered in the rf coil and from a guayule sample. It is well known that the signal-to-noise in a pulsed NMR experiment is related to a number of parameters, including the "Q" of the coil (Farrar and Becker 1971). From the above samples, the Q was different because of, among other factors, the close spacing of the "spacers" used to position the rubber disc; i.e., the enrivonment of the two samples within the rf coil was different. Also, the D_2O added to the sample would be expected to change the Q of the coil.

Two methods can be used to avoid this problem: (1) all rubber pieces used for the calibration can be centered in the middle of a "filler", e.g., a guayule sample that is known to have all the rubber extracted; or (2) one rubber sample can be scanned both with and without the "no-rubber" guayule filler, and a cor-

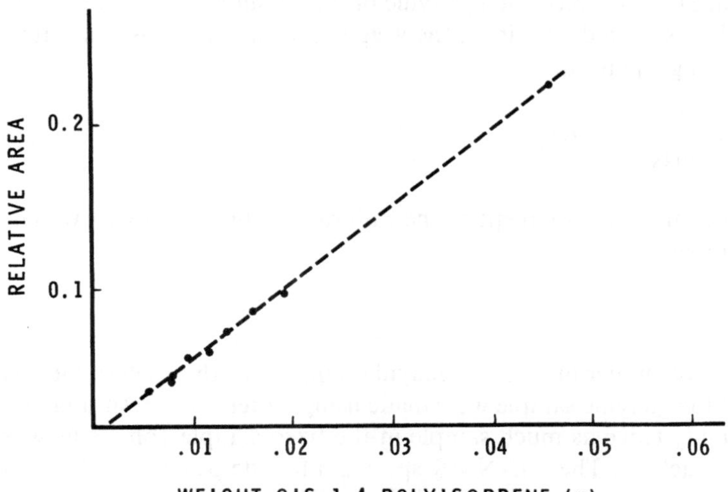

Fig. 9. Area of the methine olefinic carbon of cis-1,4-PI vs. the weight of the cis-1,4-PI sample

Table 2. Rubber weight calibration curve data

Disc no.[a]	Weight (g)	Rel. area of peak B	Ratio ($\times 10^4$)
1	0.0058	35	1.7
2	0.0085	49	1.7
3	0.0090	54	1.7
4	0.0095	63	1.5
5	0.0123	76	1.6
6	0.0136	93	1.5
7	0.0163	110	1.5
8	0.0197	131	1.5

[a] Disc made from a non-carbon-black cured rubber sheet, 1 mm in thickness, containing 93.5% rubber

rection factor obtained. Method (1) was used for the results given in this report, but we have also successfully used Method (2).

The calibration curve (Fig. 9) should be checked regularly using a known rubber sample, in order to verify that the spectrometer gain has not changed appreciably. Because of the linear nature of the calibration curve, the origin and the result of a measurement obtained from one known rubber sample are sufficient to describe a new curve.

3.3 Quantitative Results

A typical spectrum of a 0.2-g piece of a guayule branch is shown in Fig. 10. After the area of peak B has been determined, the weight percent rubber is calculated by using the following equation:

$$\text{Wt \% rubber} = \frac{B - I}{m\,(wt)} \times 100\,, \tag{1}$$

where B = area of peak B, I = intercept of the calibration curve, and wt = weight of the guayule sample.

3.3.1 Milled Bush

The weight percent of rubber in a milled guayule sample was thus determined as follows. Pellets of the guayule sample were made using a press. The resulting pellet contained roughly twice as much sample in the fixed sample volume as was obtained by hand packing. The ^{13}C-NMR spectrum for the pellet was obtained using the experimental conditions described above. The area due to peak B was measured. From Eq. (1) the weight percent of rubber in the sample was calculated.

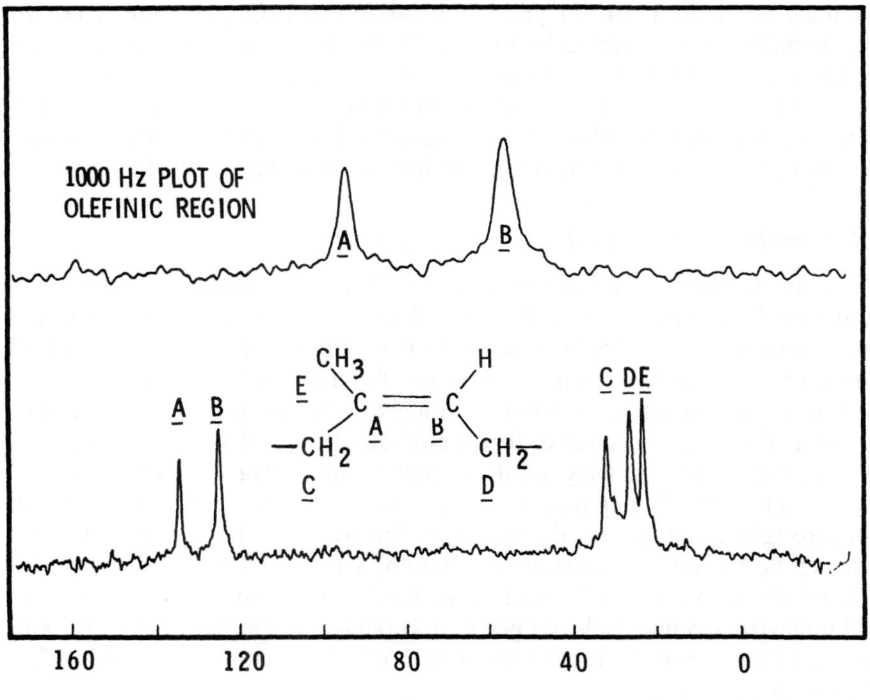

Fig. 10. 20 MHz ^{13}C-NMR sample of a portion of a branch from a guayule bush

The results for various samples are summarized in Table 3. Samples A, B, C, and D were all prepared from the same bush but represent different stages of processing. The initial rubber content of this bush was 15 ± 1.0% rubber by weight. After extraction via 11 passes with cyclohexane, the rubber content was 8.6%. Because it was necessary to physically crack the walls of cells in the bush in order for the rubber to be extractable (Nat Acad Sci 1977), this partial extraction of the

Table 3. NMR results on selected guayule samples

Sample	Description	Wt % rubber
A	Milled bush	14.0
B	Sample A with resin removed by acetone extraction	14.9
C	Sample B after two passes with cyclohexane	15.8
D	Sample B after 11 passes with cyclohexane	8.6
E	Bush with estimated 10% rubber	8.2
F	Chunk of a stem (run whole)	7.6
G	Residue from attempt at total extraction	< 1.0

rubber must be attributed to incomplete breakdown of the plant cell structure during the milling of this particular sample. Normally the rubber content can be reduced to less than 1.0% by 11 extractions, as was the case for sample G. Hence, an important aspect of this NMR method was that it worked whether or not the extraction removed all the rubber. It provides a method of measuring the residual rubber after extraction, and can be used to monitor the extraction efficiency.

3.3.1.1 Precision of the Method

The experimental precision due just to the NMR measurement, not taking into account sample preparation, was determined to be $\pm 3\%$ relative by taking successive runs on the same sample. Thus at the 15% rubber content level this would lead to an error of $\pm 0.4\%$ absolute. However, different samples of the same material, even though thoroughly mixed, do not give reproducible results within this $\pm 3\%$ error. For example, three different portions of sample B after mixing gave 15.2, 14.2, and 14.9% by weight rubber (giving an average of $14.8 \pm 0.6\%$ by weight rubber). This discrepancy was attributed to nonuniform distribution of rubber throughout the processed sample (see Nat Acad Sci Report 1977). We discuss the actual reason for this rubber variation below. Therefore, it is estimated that the precision due to all effects of the method is $\pm 1\%$ absolute. Results from several nonextracted materials are given in Table 4 and comparisons made with the results of the extraction procedure. In each case the results agree within 1.2% absolute of the NMR data.

By signal averaging for longer periods of time, the detection limit of the method can be decreased. For instance, accumulating 36,000 scans (2 h accumulation time) lowered the detection limit to 0.3% by weight rubber.

3.3.2 Guayule Branches

Segments of branches were supplied from 18 different guayule bushes. These small branches were cut directly from the bushes, and were typically 4 cm long and from 2 to 3 mm in diameter. The bark was intact on each branch.

Table 4. Comparison of weight percentages as determined by NMR and by extraction

Sample designation	Wt %	
	NMR	Extraction
H	5.9	7.1
H extracted	<0.5[a]	
I	2.3	3.3
I extracted	<1.0	
J	4.5	5.2
J extracted	<0.5[a]	

[a] These samples were run under the normal instrument conditions except 36,000 scans were accumulated

Table 5. Rubber content of various guayule branches. NMR results have an uncertainty of ±0.5% absolute

Sample	Weight % rubber		
	NMR[a]	Extraction	Difference
A-15	6.5; 5.2	3.7 (2)	2.2
A-19	3.4	3.6	−0.2
A-30	7.8	4.0	3.8
A-32	8.3; 7.0	5.7	2.0
A-33	7.1	3.8	3.3
A-35	6.8	4.7	2.1
A-36	9.8; 9.2	7.8[b]	1.7
A-38	5.1; 6.7	3.9	2.0
A-39	4.2; 4.4	2.0[b]	2.3
A-40	1.9; 3.5	5.0	−2.3
A-41	≦1.5	3.5[b]	−2.5
A-42	8.1	7.6[b]	0.5
A-43	4.9; 5.6	7.5	−2.2
A-46	2.3; ≦1.5	5.0[b]	−3.3
A-47	5.6; 4.1	6.3	−1.4
A-49	4.4	5.0[b]	−0.6
A-70	1.7	4.3	−2.6
A-71	6.7; 6.0	4.2	−2.2
		Average	0.2

[a] Second trial for selected samples was run using a different segment of the same branch. The results were averaged when obtaining the "difference"

[b] NMR analyzed for residual rubber in the materials after extraction of these samples, and found ≦0.5 wt % rubber

Lengths of 8 mm of several pieces of a branch could be run at the same time. However, to improve the filling factor, each branch was cut into seven or eight small pieces. Approximately 0.2 g of these pieces was placed in the 8-mm die (Fig. 1) and compressed using a press. In this way, a maximum amount of sample could be placed in the limited available volume.

The results for all 18 branches are given in Table 5. Each NMR sample was clipped from one of these branches. The extraction results were obtained from *several* branches of each guayule bush. The portions for the extraction procedure averaged 7.2 g.

Differences between the two methods ranged from +3.8% for sample A-30 to −3.3% for sample A-46. These results suggested that the rubber was not uniformly distributed throughout the branches. As we have seen, sometimes the extraction procedure is not complete; i.e., the rubber is not fully extracted. In order to check this possible source of error, seven of the remains after extraction were examined by NMR. In each case, the rubber content was determined to be ≦0.5% (the lower limit of detection).

It was noted that the *average* difference between the NMR results and the extraction results was +0.2% (Table 5). This suggested that the results determined via each method correctly reflected the rubber content of the sample that was analyzed. This led to the empirical conclusion that the rubber varied by approximately $\pm 3.5\%$ absolute throughout the branches. Note from Table 5 that the rubber content of several segments of the same branch, as determined by NMR, could vary by $\pm 1\%$. This variation could possibly be due to "localized" distributional differences of rubber in a branch.

Since one segment of a branch does not necessarily reflect accurately the rubber content of a bush, two or more samples per bush should be analyzed. However, one trial can certainly discriminate whether the rubber content is in the high ($>12\%$), medium (6–12%), or low ($<6\%$) range.

When segments of several of these branches were milled, the NMR results did not reflect the $\pm 3.5\%$ variation described above. This was most likely due to the mixing (homogenization) of the samples. Hence, one NMR run on a portion of a *milled* sample should be used to determine the average content from several branches.

3.3.3 Branches of Different Sizes; Bark

An entire branch from a guayule bush was cut from its trunk. Portions of this branch were examined in detail using the ^{13}C-NMR method. A sketch of the

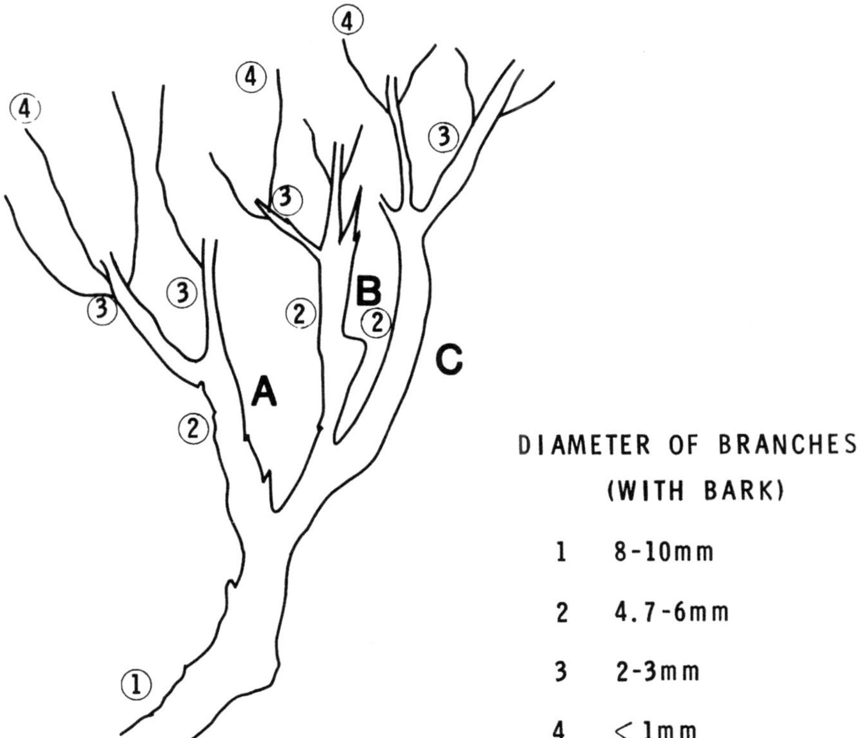

Fig. 11. Schematic of a branch cut from the trunk of a guayule bush. Not all of the smaller branches are shown in this sketch

Table 6. Branch segments from same guayule bush

Branch segment	Diameter (with bark)	Wt % bark[a]	Sample[b]	Wt % rubber			
				Bark	Branch[c]	Weighted average (branch and bark)	Average
1	8–10 mm	16	1	14	31; 28	27	27
2	4.5–6 mm	23	2A	12	26; 30	24	23
			2B	7	26	21	
			2C	10	29; 28	24	
3	2–3 mm	32	3A and 3B (combined)	7	21	16	14
			3C	8	15; 14	12	
4	<1 mm	–	4	–	–	≦0.5	–

[a] The weight % bark is from the average of several segments of the same size branches
[b] A, B, and C refer to different branches of similar sizes (see Fig. 11)
[c] Several measurements were made in duplicate

branch, showing the different segments, is shown in Fig. 11. Because it was expected that the bark may be one of the reasons for the apparent rubber variation in a bush, the weight % bark of each segment of the bush was measured. These results, and the size of each segment, are given in Table 6.

The weight % rubber was determined for the branch segments without the bark. The % rubber in the bark from each segment was independently measured. These results are also given in Table 6.

Several observations can now be made:
- From a single branch, the rubber contents of similarly sized pieces are relatively constant.
- The bark has significantly less rubber by weight % than the rest of a branch.
- Larger branches have a higher proportion of rubber than smaller branches.

In the literature, a possible source of confusion is in the use of the word "bark". We use "bark" to mean only the *outer* bark, and not the cortex (sometimes called the inner bark). Most of the rubber is present in the cortex.

The results given in Table 6 cannot be used to imply that the bush from which that branch was taken contained ≧20% rubber. The rest of the bush must also be analyzed. This is done in the following experiment.

3.3.4 Branches, Trunk, Roots

Branches of varying sizes, the trunk, and the roots of three different bushes were examined, and the results given in Table 7. These results indicate that there is little rubber in the roots and main trunk of the guayule bush, and that the rubber is concentrated in the branches. This is consistent with the previous results. For example, whenever milled samples of the entire bush were studied, the NMR and

Table 7. Rubber contents in branches of varying sizes, the trunk, and the roots of three different bushes

Bush no.	Wt % rubber			
	(5–7 mm Diameter)	(2–3 mm Diameter)	Trunk	Roots
1	3.5	2.5	<1	<1
2	7.0	4.5	≦3	<1
3	4.5	2.3	<1	<1

extraction results were in good agreement. As another example, the previous section showed that the NMR results of branches from 18 different guayule bushes agreed quite well with extraction results, which were done *only* on other branches, and not the entire bush.

The rubber content of the branches is therefore not representative of the rubber content from an entire bush. Extraction results from a "whole" bush would reflect the "dilution" effects of the trunk or roots (the portions of the bush with the least rubber.

3.3.4.1 Extraction Options

Current long-range planning calls for the rubber to be extracted from the entire bush. Based on the ^{13}C-NMR information, several new options present themselves. These options are based on extracting the rubber from the branches only, which can be separated from the plant's trunk and roots, resulting in more efficient rubber extraction. The bagasse can possibly be burned as fuel. Also possible is the harvesting of some of the branches from each bush, leaving the rest intact. If the bush continues to flourish, harvesting of the branches can continue on the *same* bush. All these options are based on the increased rubber-extraction efficiency achieved by eliminating the portions of the bush that contain relatively little rubber.

4 Conclusions

We have demonstrated the feasibility of the ^{13}C-NMR method in determining the rubber content of samples taken from guayule bushes. Each spectrum is obtained in 20 min. Results have been given for samples in the weight % rubber range of 0 to 30, with an experimental accuracy of $\pm 1.5\%$ absolute. The lower limit of detection, currently 1%, can be improved by running the experiment for longer periods of time. A 2-h scan reduces the detection limit to 0.3%.

Use of the ^{13}C-NMR method has given a picture of the distribution of rubber in the guayule bush. The rubber is mainly concentrated in the branches of the bush, with little present in the trunk or roots of the bush. For the bushes we studied, we have empirically determined that the rubber content of guayule *branches* could vary by ±3.5% absolute by weight.

Current plans for extracting the rubber from the entire bush should be reconsidered in light of these results. Our measurements of the rubber contents in the *trunks* represent a very small sampling, and further work is needed here. Currently, running a sample of several branches that are milled together is the best way to determine the weight % rubber available from all the branches of the bush.

Appendix A ^{13}C-NMR Method

A.1 Scope

This method is applicable to all types of guayule bush samples; e.g., a solid chunk of bush which in size is ≦8 mm long and 8 mm in diameter, a bush after milling, bush with resin removed, or bush at any level of resin extraction.

It requires about 15 min to get the sample in pellet form, 20 min of instrument time, and another 15 min for determining areas and calculations.

Precision of the method, for the 20-min averaging time, is ±1.5% absolute, and holds for any rubber content. The lower limit of detection can be reduced to 0.3% by weight rubber by accumulating for 2 h.

A.2 Principle of the Method

The ^{13}C-NMR peak of interest is "peak B". This is the peak area due to the methine olefinic carbon (at approximately 126 ppm).

The peak B area in the unknown guayule sample is compared to the peak B areas of known rubber samples used as calibration standards, and the weight of rubber in the guayule sample is thus determined.

A.3 Apparatus

A standard, easily available NMR spectrometer, e.g., a 20 MHz ^{13}C Fourier transform spectrometer equipped with a 10-mm probe, and preferably with an external lock (to remove the need for using D_2O which may "swell" the sample).

10-mm NMR sample tubes.

A pellet press capable of making pellets 7–8 mm in diameter, and 7–8 mm long.

A.4 Reagents

D_2O of at least 90% deuteration (not needed if the spectrometer includes an external field lock).

A.5 Calibration

Obtain one or more gum rubber standards, preferably ≧90% rubber (cis-1,4-polyisoprene). Make two or three samples of varying weight; this range of weights should cover the estimated range of the rubber weights in the guayule samples. The rubber sample should be centered in a filler of rubber-extracted guayule in order to make the Q of the rf coil as nearly the same as when the guayule samples are run. Position each sample between two spacers in a 10-mm NMR tube such that it is approximately in the center of the rf coil. Add D_2O for a lock signal. Obtain the NMR spectrum according to the experimental conditions given in the following list.

Using the electronic integrator of the spectrometer, measure the peak B area for each sample. Plot a calibration curve, with the peak area along the ordinate and the weight of the rubber sample (corrected for not being 100% rubber, if necessary) along the abscissa. The points should fall along a straight line, with the intercept at the origin (see Fig. 9).

Obtain a linear least-squares fit of the calibration data, and use it to calculate the weight percent rubber from area B (see A.7).

The calibration curve should be checked with a known rubber standard regularly. Any significant deviation from the calibration curve, signifying a change in the overall gain of the spectrometer, would necessitate obtaining a new calibration curve.

A.6 Procedure

Make up the guayule samples by placing approximately 0.3 g of the sample into the pellet maker. If a chopped bush, unmilled, is being examined, it may be run without treatment. The aim is to get as compact a sample as possible. The maximum size of the sample should be 8 mm in diameter and 8 mm in length, taking into account any swelling due to D_2O (the volume of the sample typically increases by about 20%). All samples are to be placed totally within the volume of the NMR rf coil, and thus quantitative conditions are assured throughout the sample.

Place a vortex plug in the bottom of a 10-mm tube, add D_2O to cover about 3 mm over the top of the vortex plug, place the guayule pellet on top of the vortex plug in the tube, and place another vortex plug on top of the guayule pellet (see Fig. 2b). The D_2O level should at least come to the top of the upper vortex plug.

The vortex plugs should have their vanes stripped away, in order to increase the amount of D_2O near the rf coil. If an external lock is available, one can omit the addition of the D_2O, and thereby increase the size of the guayule pellet.

Scan the sample using the following typical instrument conditions:

Nucleus	^{13}C
Filter bandwidth	4000 Hz
Spectral width	4000 Hz
Number of transients	6000
Acquisition time	0.2 s
Pulse width	90 pulse
Pulse delay	0
Weighting factor	−0.05
Fourier number	4096

Use proton broadband decoupling

Plot expanded scale spectrum of the olefinic region, and obtain the area of peak B using electronic integration.

A.7 Calculations

Determine the slope and intercept of the calibration curve. Thus,

$$\text{Wt \% rubber in sample} = \frac{B - I}{(M)(Wt)} \times 100,$$

where

B = area of peak B
I = intercept of the calibration curve
M = slope of the calibration curve
Wt = weight of the guayule sample

References

Banigan TF, Verbiscar AJ, Oda TA (1981) Guayule rubber analysis method comparison. Phoenix Rubber Meeting, Phoenix, AZ, October 1981

Breitmaier E, Voelter W (1974) ^{13}C-NMR spectroscopy. Verlag Chemie, Weinheim, p 38, pp 45–46, 51–54

Farrar TC, Becker ED (1971) Pulse and fourier transform NMR. Academic Press, New York, p 39

Gowans WJ, Clark FE (1952) Determination of rubber hydrocarbon by a modified bromination method. Anal Chem 24:529–533

Gutierrez R, Ransaw H (1980) Rubber content of the Guayule Bush using attenuated total reflectance infrared spectroscopy. Third Int Guayule Conf, April 27–May 1, 1980. Pasadena, CA

Hanson G, Perry D (1981) Turbidimetric method of rubber analysis, included in Banigan et al. 3

Hayman E, Yokoyama H, Schuster R (1982) Carbon-13 nuclear magnetic resonance determination of rubber in Guayule. J Agric Food Chem 30:399–401

National Academy of Sciences (Report of an Ad Hoc Panel of the Board on Agriculture and Renewable Resources, Commission on Natural Resources; and Advisory Committee on Technology Innovation of the Board on Science and Technology for International Development, Commission on International Relations) (1977) Guayule: An Alternative Source of Natural Rubber, Washington, DC, p 35, pp 43–47
Schaefer J (1972) Comparisons of the carbon-13 nuclear magnetic resonance spectra of some solid cis- and trans-polyisoprenes. Macromolecules 5:427–440
Visintainer J, Beebe DH, Myers JW, Hirst RC (1981) Determination of rubber content in Guayule Bushes by carbon-13 nuclear magnetic resonance spectrometry. Anal Chem 53:1570–1572

Nuclear Magnetic Resonance and Pollen Quality

C. Kerhoas and C. Dumas

1 What is the Pollen Grain?

1.1 Organization and Role

The pollen grain represents a biological entity which assures the dissemination of sperm cells, or their progenitors. Each grain is surrounded by a complex wall which has both mechanical protective properties and a sophisticated physiological role during the pollen-pistil recognition events mediated by specific proteins and glycoproteins (review Clarke and Gleeson 1981). Some pollen species may induce hay fever or seasonal asthma in sensitive persons. It has been demonstrated that less than 2 mg of grass pollen are sufficient to induce the asthmatic response of a single flowering season. The reaction is provoked, not by the grains themselves, but by the glycoproteins and proteins located in their walls (Howlett and Knox 1984).

Table 1. Proteins in intine and exine sites of pollen grains. (After Knox et al. 1975)

Protein	Present in	
	Intine	Exine
Dehydrogenases		
NADH and succinic dehydrogenase	–	+
Oxidases		
Cytochrome oxidase	–	+
Transferases		
Phosphorylase	(Rapidly diffusible)	
Ribonuclease	+	–
Hydrolases		
Acid phosphatase	+	–
Amylase	+	+
Cellulase (β-1,4-glucanase)	(Rapidly diffusible)	
Esterase	+	+
Invertase (β-fructofuranosidase)	(External to plasmalemma)	
Polygalacturonase (Pectinase)	(Rapidly diffusible)	
Protease	+	+
Allergens		
Ragweed antigen E	+	+

Table 2. Average time of earliest detectable emission from exine and intine sites in pollen grains of different structural types. Protein emission detected by the Coomassie blue stain-fixing procedure and the "pollen print" method. (Heslop-Harrison et al. 1975)

Species and pollen type	Emission of sporophytic fraction from sexine sites	Emission of gametophytic fraction from intine sites
Alopecurus pratensis; *Dactylis glomerata*, Gramineae (tectate; single operculate aperture; smooth exine)	<25 sec	2–3 min
Silene vulgaris, Caryophyllaceae (tectate; many operculate apertures; moderately smooth exine)	<30 sec	1–5 min
Cosmos bipinnatus, Compositae (tectate; three colpi; spinly exine)	2–5 sec	10–20 min
Ambrosia trifida, Compositae (tectate; three colpi; exine with low spines)	2–5 sec	co. 30 sec
Iberis sempervirens Cruciferae (murate; three colpi)	<20 sec	6–10 min
Hibiscus rosa-sinensis Malvaceae (tectate; many nonoperculate apertures; moderately smooth exine)	<30 sec	4–5 min

The pollen wall is also the site of initial contact with the stigma surface in male-female recognition. What potential has the wall in recognition? It contains extracellular enzymes. Initially several works showed that the wall proteins occurred in two quite different domains (Heslop-Harrison 1975; Knox et al. 1975):

- diploid parentally specified proteins held in a labyrinth of arcades in the outer wall, the exine (sporophytic domain).
- haploid-specified proteins are laid down in the inner wall layer, the intine (gametophytic domain).

A considerable number of proteins and glycoproteins (proteins with sugar chains attached covalently) are present in the pollen wall (see Table 1). The proteins housed in the exine are the first to make contact with the pistil, while those in the intine may be released more gradually during germination (see Table 2). There is thus evidence for a considerable number of potential informational molecules at the surface of the male partner. There is now biochemical and immunological evidence that many are specially synthesized in the pollen grain, while others are common with those of somatic tissues of the same plant, e.g., *Gladiolus* (Clarke et al. 1977), *Prunus* (Raff et al. 1980; Mau et al. 1982) and *Zea* (Porter 1981).

On the other hand, pollen grains are quite fascinating for biologists because each grain provides an accessible, isolated, intact multicellular organism. About two-thirds of pollen grains are bicellular, carrying a large vegetative cell and the progenitor cell of the male gametes included within (Fig. 1) (e.g., fruit trees be-

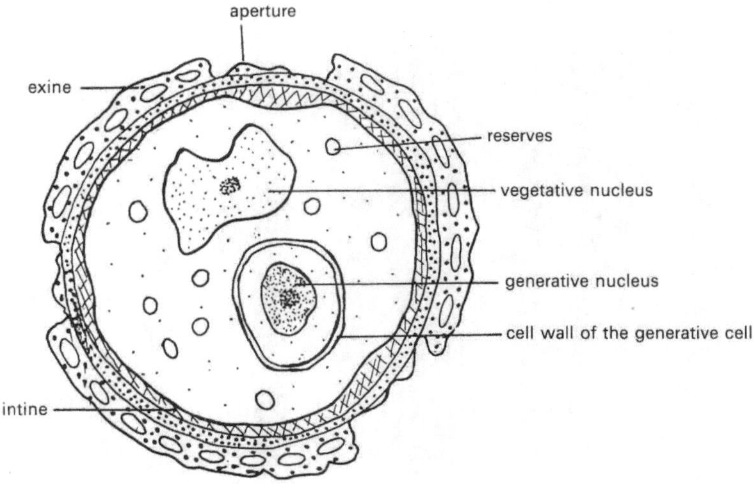

Fig. 1. Bicellular pollen grain

longing to Rosaceae, clovers other lucernes in Leguminoseae, tomato, tobacco, potato in Solanaceae, etc.).

The remainder are tricellulae, and the pair of sperm cells are borne directly within the vegetative cell from the progenitor cell after a special mitosis (Fig. 2).

Recently, several authors demonstrated that the two sperm cells are linked together and associated with the vegetative nucleus (Russell 1983; Dumas et al. 1985). We termed this peculiar association male germ unit (Fig. 3) (Dumas et al. 1984a). In addition, both sperm cells are not equivalent. Thus, double fertilization does not involve random fusions of a pair of sperm cells and female reproductive cells (female germ unit, see Dumas et al. 1984d). In the following section we will discuss this important point according to the concept of pollen quality.

1.2 Water Content, Membrane Organization and Pollen Viability

The pollen grain is virtually a dehydrated organism. The water content at time of dispersal varies widely among different pollen types (Table 3). This dehydration stage is the process by which pollen grains become dormant at maturity before reactivation on a compatible hydrated stigma. To prepare for this dormant period, the developing pollen grains become greatly modified morphologically, chemically, and physiologically while inside the anthers. However, although such modifications as dehydration may be highly protective and therefore of survival value, they are not necessarily a prerequisite for resistance to adverse conditions occurring during pollen transport for pollination. In fact, pollen grains do not attain a truly dormant state (dormancy is primarily a stress-avoidance mechanism), but rather become more or less well adapted for surviving water stress (Dumas 1984).

The water content of living pollen grains varies considerably among different families, with most recorded values lying between 15 and 35% of fresh weight at

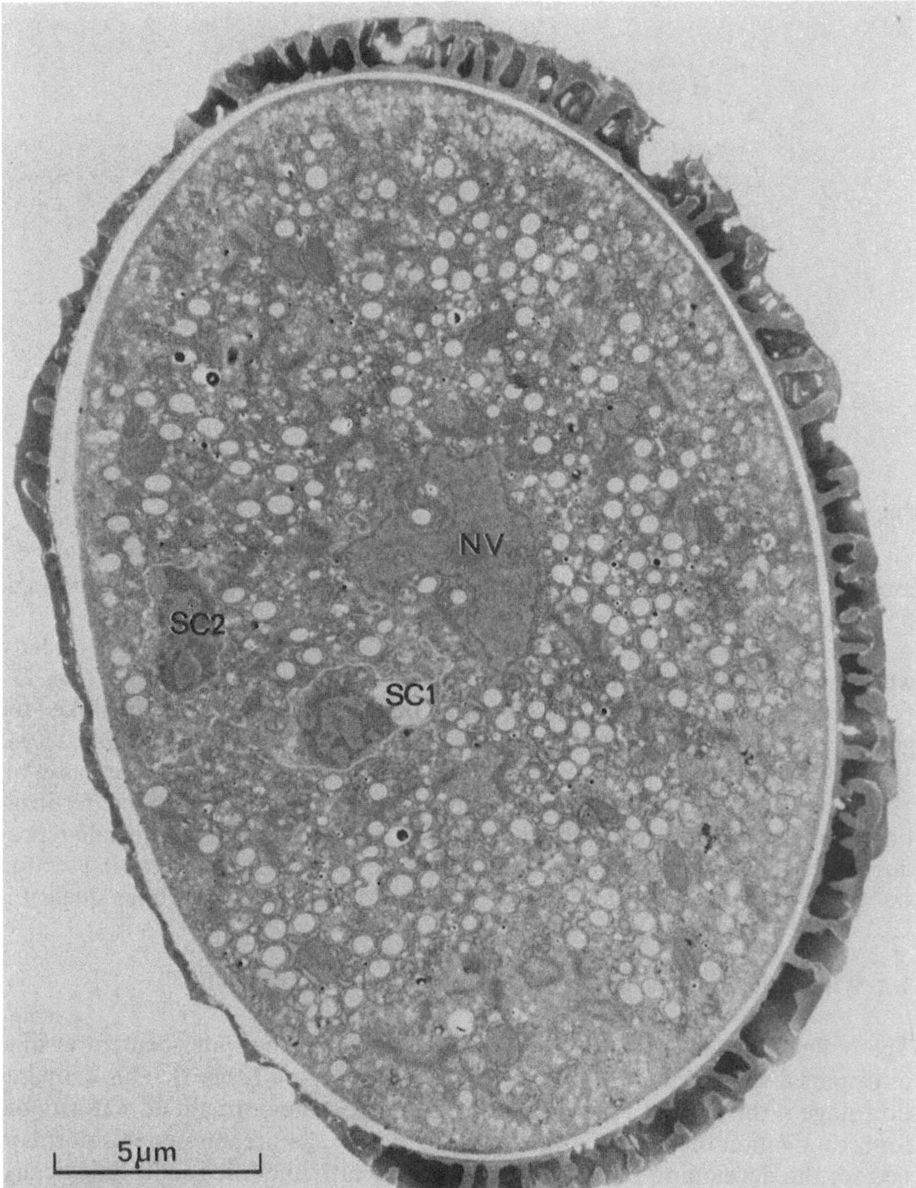

Fig. 2. Tricellular pollen grain, e.g., *Brassica*. *VN* vegetative nucleus, *SC 1* sperm cell 1, *SC 2* sperm cell 2. ×6000

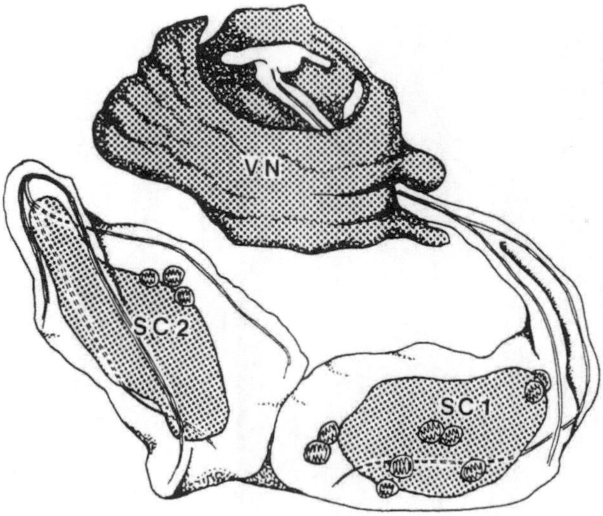

Fig. 3. Schematic diagram showing the arrangement of nuclei and microtubular arays in the two sperms (*SC 1*) and (*SC 2*). (Mc Conchie et al. 1984)

the time of dispersal (see reviews, Stanley and Linskens 1974; Knox 1984a). We have noted some extreme values of 6% in poplar pollen during storage in the refrigerator, and 60% in *Zea mays L.* (Table 3).

In the literature, little information is available on pollen longevity, generally estimated directly by the ability to get the seed. In our laboratory we noted some significant differences according to the genotype, the relative humidity (RH) value and obviously the conditions of storage. For example, at room temperature of 20 °C and 60% of RH, pollen longevity is around 2 h in corn and in wheat.

Also, there is no water content expenditure in viable pollen under isothermic conditions, suggesting that water content is regulated within certain limiting temperatures commonly experienced by the pollen. This characteristic is perhaps mediated by the plasma membrane or the wall of the pollen grain with its viscous pollen coat. This may be an important adaptation for pollen survival during dispersal.

Table 3. Average of water content in some pollen species determined by gravimetric analysis

Species	Water content (%)	Longevity	Reference
Corn	57–60	1–2 h	
Wheat	55	1–2 h	
Triticale	55	1–2 h	
Pumpkin	47	1 day	
Oil rape	18–20	1 week	Dumas et al. (1983)
Typha	17		Hoekstra (1984)
Poplar	9	2–3 days	Hamilton (1976)

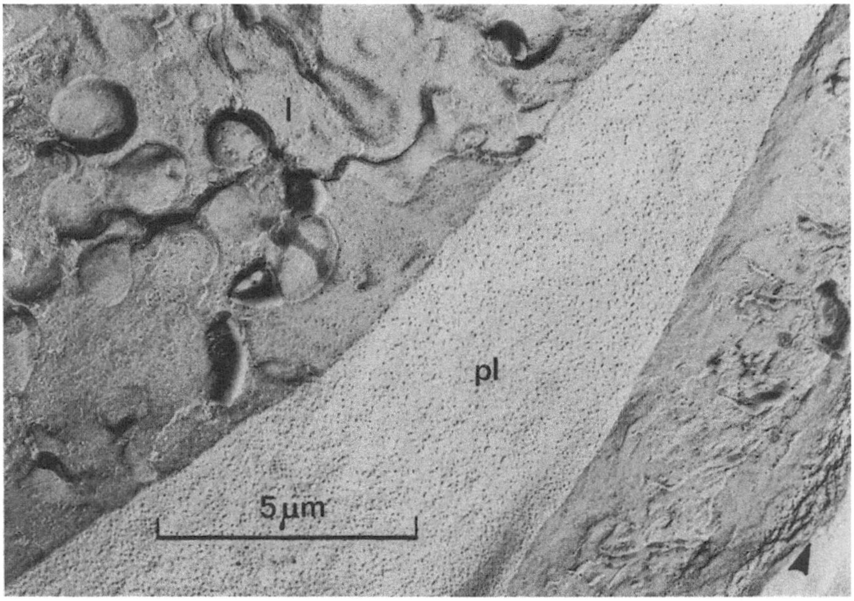

Fig. 4. Freeze fracture of *Brassica* pollen wall showing the plasma membrane of the vegetative cell with a typical particulate structure characteristic of biological membranes. *I* Intine; *pl* plasmalemma; *arrowhead* direction of the shadowing. (Dumas et al. 1984a) ×84,000

It is very interesting to note the close correlation between pollen longevity or viability and water content.

A fraction of liquid water (vital water) may be eliminated through the porous plasma membrane of dead grains (Burke et al. 1976), but in a living grain the plasma membrane is an effective barrier. What is the state of the plasma membrane of mature pollen? Three types are known:

- those with a porous and ineffective plasma membrane, associated generally with a high water content and short viable life, e.g., *Secale* pollen (Heslop-Harrison 1979a, b);
- those with a continuous and effective plasma membrane, associated generally with an intermediate level of water content and longer viable life, e.g., *Brassica* pollen (Dumas and Gaude 1983) (Fig. 4);
- those with a discontinuous plasma membrane and a low water content (less than 10%), e.g., *Populus* pollen (Fig. 5).

In the second case, the typical particulate structure of characteristic biological membranes have been detected in the plasma membrane by freeze-fracture (see Fig. 4 and Dumas et al. 1984a).

Both the first and third types belong to plants whose pollen is mainly dispersed in air currents; the grain wall (exinic layer) is generally thin with pollencoat reduced or absent. In the second type, the pollen is generally dispersed by specific animal vectors, and the wall is thick and reinforced, with a copious covering of sticky pollencoat (e.g., *Brassica*). Differences in the chemistry of the pollencoat are suggested by the ultrastructural studies of Hesse (1979, 1980), who has de-

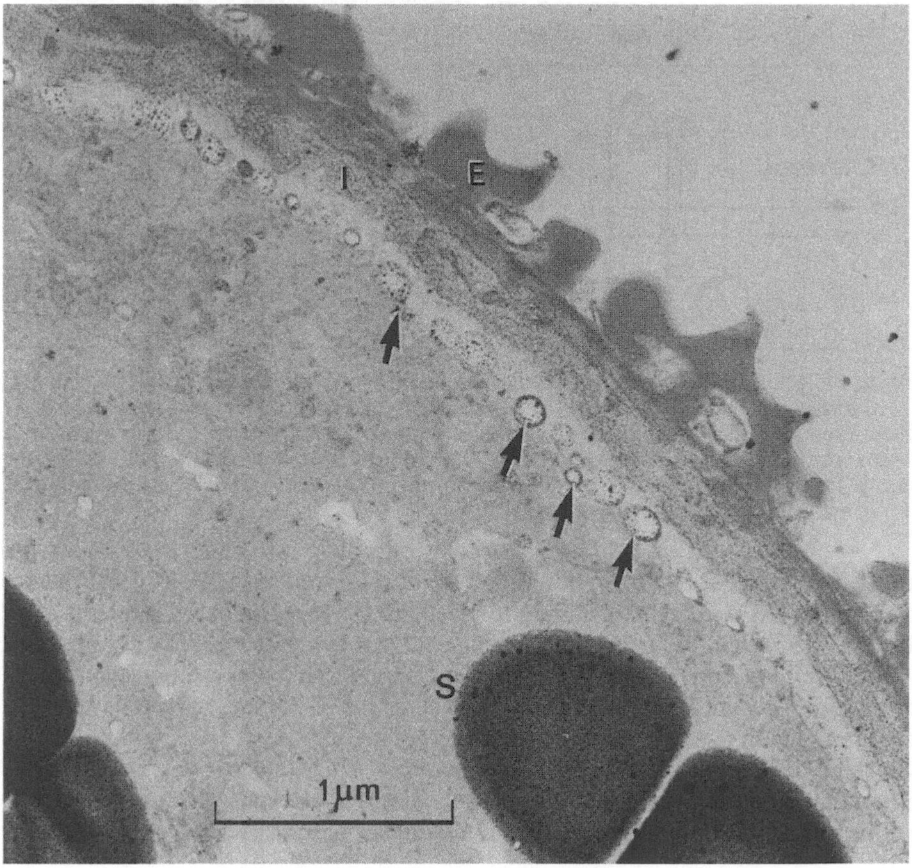

Fig. 5. Vesicular aspect of the plasma membrane (*arrows*) in a low water content type of pollen after PATAg treatment (e.g., *Populus nigra*) *I* intine; *E* exine; *S* starch. (Gaget M, unpublished data, and CMEABG) ×32,000

tected "active" and "inactive" types, distinguished by their electron density and heterogeneity, and site at the exine surface.

Today, the most interesting destructive technique available to appreciate pollen viability is a cytological technique called fluorochromatic reaction (FCR). This technique, adapted to pollen by J. Heslop-Harrison and Y. Heslop-Harrison in 1970 is based on the accumulation of a fluorescein dye within the pollen grain. An explanation is suggested for this intracellular accumulation: that the nonpolar substrate passes inward through the plasma membrane more rapidly that the free fluorescein liberated after hydrolysis by a cytoplasmic esterase passes outward (Rotman and Papermaster 1966). This technique has been correlated with the in vitro tests of germinability and was found to be highly significant (Y. Heslop-Harrison and Shivanna 1977). The FCR is a test principally of membrane integrity (J. Heslop-Harrison et al. 1984) and a negative score cannot be interpreted as an inviability state for the pollen (Dumas et al. 1984b) (Fig. 6). In vitro culture methods involve germination of pollen on an agar medium containing calcium

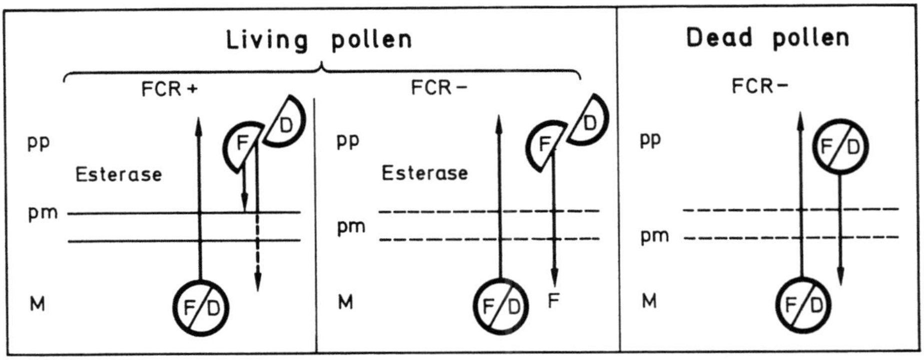

Fig. 6. FCR test and plasma membrane permeability. (After Dumas et al. 1984 b) *F/D* fluoresceine diacetate; *D* diacetate; *F* fluoresceine; *M* extracellular medium; *pm* plasma membrane; *pp* pollen protoplasm. The regulation of the water content (WC) involves both temperature effect (t°) and duration (T). Thus we can formulate this mechanism by the simple following formula: WC regulation = f (t°) + f (T) + f (t°,T). The last parameter f (t°,T) may characterize the membrane

and boron (semi-solid) or in liquid culture (see Knox 1984 a, b). Unfortunately, this simple test is applicable mainly to bicellular pollen types (e.g., Williams et al. 1982), as most tricellular pollen types show a very low percentage of germination in vitro, except corn pollen, for example. These latter types generally require a quite sophisticated medium (e.g., Roberts et al. 1983). All these previous tests, including of course direct in vivo experiments, involve loss of the sample tested, which is no longer available for subsequent breeding tests or other experiments. A new method developed by Dumas et al. (1983) overcomes this problem. It depends on the use of Nuclear Magnetic Resonance Spectroscopy (NMR).

2 ^{1}H-NMR – Water Content

2.1 Method

This method provides data that are rich in structural as well as dynamic information. It utilizes the principle of resonance of the hydrogen atomic nucleus, ^{1}H, when placed in an intense magnetic field. Atomic nuclei oriented by a strong magnetic field absorb radiation at characteristic frequences. The parameters that can be measured on the resulting spectral lines (line positions, intensities, line widths; multiplicities, and transients in time-dependent experiments) can be interpreted in terms of molecular structure, conformation, molecular motion, and other processes (e.g., Cohen 1983).

What information can be obtained by NMR analysis of the behavior of the proton of water molecules of the pollen grain? These particles have three basic properties – charge, mass and spin, or angular momentum. This spin and the physical constants associated with it – such as the angular momentum – cannot be measured by a direct experiment, but can be derived from the behavior of the magnetic moment associated with a spinning charge. When placed in a magnetic field, protons can be grouped according to their energy levels. ^{1}H-NMR allows us to study transitions between these states and to observe the absorption frequency due to hydrogen in water. Similarly ^{31}P-NMR allows us to investigate phosphorus in membranes (Cohen 1983).

In the liquid state, the NMR signal has a narrow line-width (approx. 1 Hertz) while with solid state (as with a dry pollen sample) a broad line-width (approx. 600 Hertz) may be obtained.

Our first experiments on a population of hydrogen nuclei from pollen grains gave a resonance around 42 Hz. The signal obtained is a shift which corresponds to a dispersion line. Its stage is derived from the Bloch equation. The area under the line is directly proportional to the hydrogen protons present.

2.2 General

Biological tissues have an extremely complex structure, which makes it difficult to elucidate the basic characteristic of cell water. Also, most of the studies on water structure are carried out with solid/aqueous systems which are much more simple than biological tissues. Chemical, physical (ESR, NMR, thermodynamic), biochemical (enzymatic activities) methods have, nevertheless, provided knowledge about some of the details of water structure (Peschel and Belouschek 1979; Hazlewood 1979). The difficulties implied by the biological material could account for controversy surrounding the physical state of intracellular water. Opinions range from the view that all the water in the cell is different from bulk water, to the opposite, in which 5–10% of water is not bulk water (Clegg 1979). In fact, there are two phases of cell water, one of which is bulk water, that is water in dilute solution (Clegg 1979), the other is surface water (Drost Hansen 1976). The properties of water adjacent to most solid interfaces differ from the corresponding bulk properties. Such "surface-modified" water is referred to as vicinal water (Etzler and Drost Hansen 1979). Water attached to an underlying hydrophylic surface is subject to a restricted molecular motion. This effect is transmitted via hydrogen bonds more or less deep into the bulk phase (Peschel and Belouschek 1979). This deep effect is subject to controversy, but most physicists conceive that electrostatic forces on macromolecular surfaces can influence water molecules over distances no greater than a few Å (Hazlewood 1979), or 25 Å for Drost Hansen (1976), Wiggins (1979). No physical model can explain a much longer effect that is called long-range order.

The degree of structural changes in vicinal water depends on whether the interface is more or less hydrophilic (Peschel and Belouschek 1979). In living systems, the surface-to-volume ratio is high; also, it must be expected that vicinal water may play an important part in cellular life. In these general data, we have to mention some ambiguous terms such as "bound", "ice-like" and "ordered" water which lack quantitative description (Hazlewood 1979). All these notions are important in understanding water behavior for an organism so complex as pollen.

2.3 Water Content Evolution

Pollen grains are collected by hand (entomogamous types) or by direct gravitation (anemogamous types) (see Stanley and Linskens 1974) and stored directly in small glass containers at controlled temperature.

For each experiment either NMR or FCR measurements were carried out simultaneously. For NMR measurement, we used the so-called solid NMR technique, in which water content analysis has been realized without any liquid medium, to avoid the proteins released leading to a loss of informational molecules (see Sect. 1). Both ^{1}H-NMR and gravimetric methodology have demonstrated

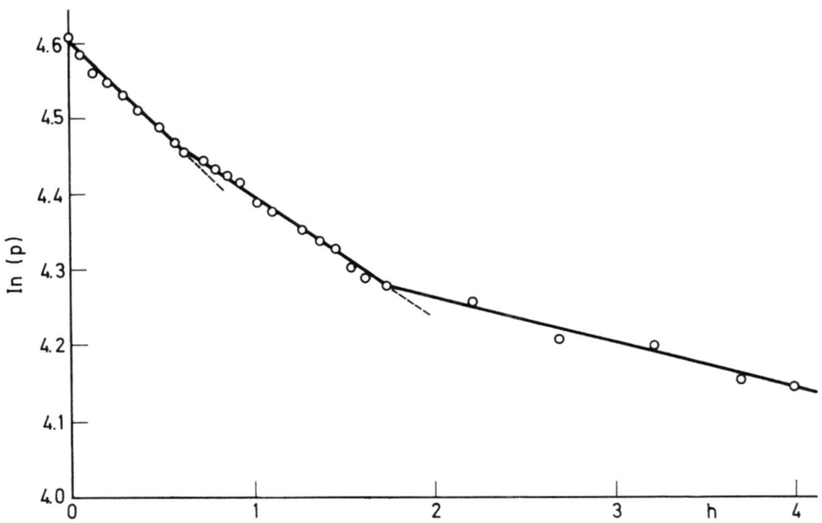

Fig. 7. Gravimetric study of pumpkin pollen dehydration at 24.5 °C and 40% RH. (Kerhoas et al., in press, 1986)

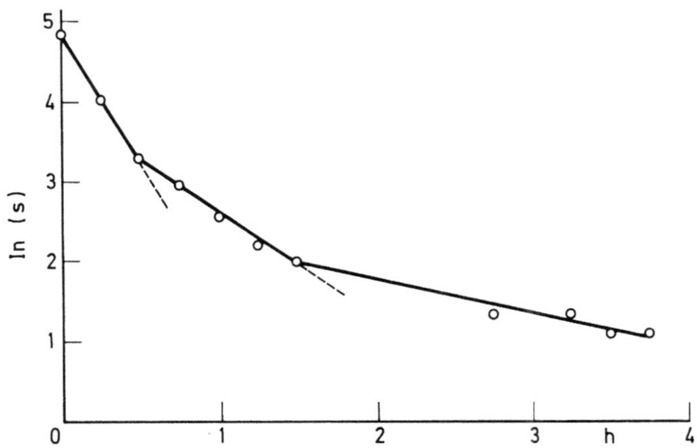

Fig. 8. Kinetics of pumpkin pollen dehydration using ^{1}H-NMR. *s* ^{1}H-NMR peak integration. (Kerhoas et al., in press, 1986)

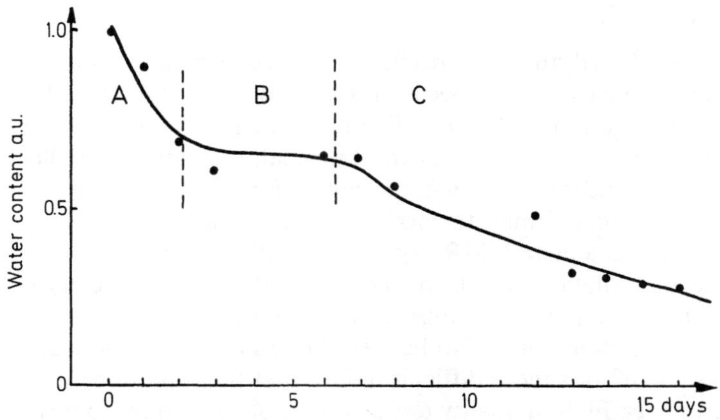

Fig. 9. Loss of water content during the pollen aging. Three steps (*A* to *C*) are visible. (*a.u.* arbitrary unit) (Dumas et al. 1983)

that during aging of *Brassica* and pumpkin pollens, water content decreased (Dumas et al. 1983) (Fig. 7 and 8). In *Brassica* pollen, NMR results were closely correlated with FCR data as markers of the loss of viability (Fig. 9). Thus, loss of viability is correlated with water content.

2.4 Water State Evolution

2.4.1 Relaxation Time

To detect water state evolution in pollen grains during aging, we carried out a second series of NMR experiments involving the nuclear relaxation phenomenon. When a radio frequency is applied to a sample, a transverse magnetization is induced. This magnetization declines when the radio frequency of excitation stops and the excited system returns to equilibrium. This relaxation phenomenon is possible if nuclei are equally dispersed on the precession cone. Transverse magnetization decays exponentially with the relaxation time T_2.

Following the Meiboon Carr Purcell spin echo method (see Duplan and Dumas 1984), we measured T_2 for several species of pollen grains during aging.

A model has been employed which includes two types of cellular water, bound water (low T_2 values) and free water (high T_2 values).

Thus, in a pollen grain population, we may represent this phenomenon by the following formula:

$$\frac{1}{T_2} = P_1 \frac{1}{T_2(b)} + (1 - P_1) \frac{1}{T_2(f)}.$$

Where

b: bound water molecules
f: free water molecules
P_1: proton spin-lattice relaxation time
T_2: proton spin–spin relaxation time

2.4.2 Hydric Compartmentation

The curves obtained with NMR and gravimetric data are correlated and suggest they are composed of three parts: three speeds of water loss, that is three hydric compartments in pollen organisms, *Brassica* (Dumas et al. 1983) and pumpkin (Kerhoas et al. in press, 1986). These studies are statistical, in relation to pollen heterogeneity, and imply some moderation in the conclusions, but the results are repetitive. What about the three compartments in pumpkin pollen?

a) First compartment: In a pulse NMR experiment, with a sequence of short pulses of radiofrequency radiation and 0.2 ms between every pulse, we can observe a decrease of relaxation time T_2 during the first hour of dehydration (Fig. 10). The first compartment could also be identified as a free one, so called "bulk water" compartment? It is more difficult to interpret the rupture between the two first compartments: FCR, a quality test described above, brings some information about this rupture. This test points to some evidence of plasmalemma integrity in peculiar during pollen senescence. FCR data are closely correlated to relaxation time T_2, especially when relaxation time is greatly decreased (Fig. 11).

We can conclude that a correlation exists between palsmalemma integrity and bulk water content. When the "bulk water compartment" disappears, plasmalemma would break down. Another author, Simon (1978) pointed out a membrane disorganization below a water content of 20%.

Other NMR experiments corroborate plasmalemma evolution during pollen aging: Relaxation time study, during a pulse NMR experiment with 2 ms between every pulse, points out, the occurrence of many T_2 for a same hydration and viability state (Fig. 11). Furthermore, it has been demonstrated that proton T_2 of water in different nonoiled seeds reflected an heterophasic behavior of seed water (Askochenskaya 1980 cited in Ratkovic et al. 1982). In a pollen fully hydrated,

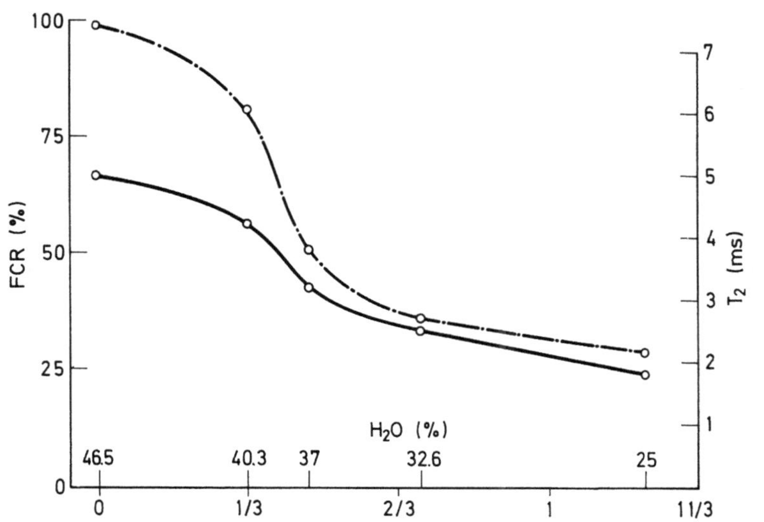

Fig. 10. ^{1}H-NMR relaxation time T_2 and FCR evolution during Pumpkin pollen dehydration. (Kerhoas et al., in press, 1986) —·— FCR; —— T_2

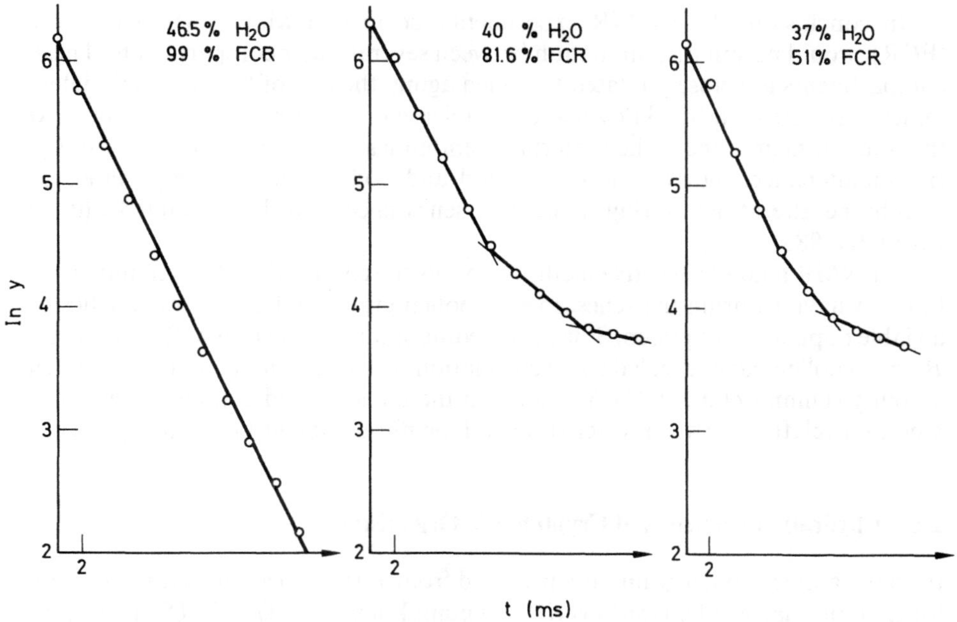

Fig. 11. Evolution of hydric compartmentation of pumpkin pollen during aging. (Kerhoas et al., in press) *In y* [1]H peak integration

no discrimination between different compartments is possible: when pollen is slightly dehydrated (40% H_2O) and less viable (81.6% FCR), three T_2 seem to appear: the third can occur for water which is more free. When pollens go on to senescence, their membrane disorganization leads to a loss of water control and a discharge of intracellular water which becomes free. When pollen viability is largely reduced (51% FCR), this free water becomes more important. This water behavior is closely related to decreasing pollen viability, rather than decreasing total water content (Fig. 10). In fact, the loss of bulk water determines the beginning of plasmalemma destructuration and the decrease of pollen viability.

These observations confirm:

- the occurrence of bulk water in plasmalemma integrity
- the occurrence of membrane integrity in a viable pollen.

In the same way, Ratkovic et al. (1982) have pointed out three T_2 components in endosperm of seeds containing lipids: a solid and a semi-solid T_2 component found in endosperm containing 8–9% water and, for a more hydrated endosperm, a liquid component appears.

b) Second compartment: This one begins to lose its water content after the plasmalemma go on destructuration. The speed of dehydration is slower. In fact, this dehydration concerns water more or less bound to plasmalemma and internal membranes. The energy which allows its liberation becomes greater, until all membranes are disorganized. This could account for the second rupture and the third compartment.

c) Third compartment: All water is freed from the dead organism.

In conclusion, these NMR experiments, correlated with a cytological test (FCR), would permit discrimination between several water compartments. These compartments are closely related to pollen aging: the loss of "bulk water" determines plasmalemma breakdown and the release of water more or less bound to the internal membrane. When internal membranes possess little water for integrity maintenance, they are destructurated and pollens die. The importance of membrane alterations during aging and senescence have been pointed out by Mazliak (1983).

^{1}H-NMR a nondestructive method, reveals the complexity of water and membrane behavior during the senescence of pollen grains and the characteristics of a viable population of pollen. Some preceeding studies with ^{1}H-NMR spectra and *Brassica* pollen have revealed some correlation between water content and pollen viability (Dumas et al. 1983). We confirm these results and complete them with a new correlation between water state and membrane integrity during aging.

2.5 Dehydration Control and Cryptobiotic Organisms

In some families pollen grains are dispersed from anthers in a fully hydrated state: for example maize pollen (56% H_2O) and pumpkin pollen (47% H_2O). These pollen are not in a dormant state and are unable to survive in adverse environmental conditions encountered outside the anther, that is drastic dehydration (see above). Also, they have a short life (1 or 2 h) if they are not rapidly in contact with receptive and compatible stigma (Dumas et al. 1984a).

Furthermore, these "tender" pollen in which water content is higher than 15–20% of fresh weight cannot survived after direct immersion in liquid nitrogen (−196 °C) without any special precautions (Levitt 1980; Kerhoas et al. 1984). Consequently, it is essential to practice dehydration before ultracooling (Barnabas and Rajki 1981); this treatment avoids the mechanical stress due to the cristallization of water. This dehydration must be slow and controlled, and implies a striking study of water compartments as described above. The key solution in storage should consist of intermediate dehydration in order to avoid dehydration stress (vital water has to be preserved) and freezing stress (freezing water has to be eliminated) (Gay et al. 1984). DSC and ^{1}H-NMR are the two methods employed in order to control such dehydration.

Clegg (1978) works on an interesting model, that is, an organism which can be severely dried and exists in a viable state without any metabolism. This original biological organization is one of a cryptobiote: *Artemia* cyst. The dried cryptobiotic organism can be "revived" by the addition of a suitable amount of water. The dried cysts can be stored for infinite periods of time at low temperature and pressure. Thus *Artemia* cyst studies could provide most interesting features on the organization of a water and freezing stress-resistant system. By using some physical parameters (isotherm, NMR parameters, dielectric and DSC measurements), the author has obtained a picture of the hydration dependence to the cyst metabolism. We have tried to establish some analogies with pumpkin pollen (Table 4). In pollen – cyst-like – the water content range is related to physical measurement transition.

Table 4. Comparison between cyst (Artemia type) and pollen (pumpkin type) from Clegg data (1978) and Kerhoas et al.

Water content expressed in %		Water state	Metabolism behavior	Organism behavior	
Cyst	Pollen			Cyst	Pollen
30	24	Vicinal water	Ametabolism	Dormant state	Dead state
30–60	24–36	Vicinal water	Reduced metabolism	Viable state	Viable state
60	37–47	Vicinal + bulk water	Normal metabolism	Viable state	High respiration rate
>60	>47	Vicinal + bulk water	Normal metabolism	Viable state	Hydration on the compatible and receptive stigma

According to Clegg, vicinal water seems to have a key importance for the metabolism. When an organism loses its cellular control (plasmalemma destructuration) and its intracellular surface organization, it cannot be revived. These phenomena occur during pollen dehydration, which appears to be a crucial event during the pollen storage

Clegg assigns a real importance to the metabolism and to the water state. It is the basic cellular matrix within which most of metabolism takes place (Clegg 1979). Vicinal water structures are generated by intracellular surfaces which are under cellular control. Also, when an organism loses its cellular control (plasmalemma break down), and its intracellular surface organization, in an irreversible way, it cannot be revived. This is how pollen behaves during dehydration (see above). Consequently, this problem for pumpkin pollen conservation is important, as we cannot dehydrate the pollen below 24% of water. Thus in order to immerse the pollen in liquid nitrogen, we have to decrease some of the amount of nonfreezable water (we have extrapolated these features from Becwar et al. 1983). The content of nonfreezable water has been determinated in pumpkin pollen by DSC (differential scanning calorimetry) and is relatively slight, i.e., 12%. Also, our object is to dehydrate pollen below 12% without modified intracellular surface organization, essential in vicinal water structure generation.

Why is the metabolism state reversible in cyst and irreversible in pumpkin pollen? We can say that this irreversible state is associated with an irreversible destructuration of the internal membranes in pollen, as suggested above. Also, it appears essential to undertake physical membrane studies.

3 ^{31}P, ^{13}C-NMR and Membrane State

3.1 Structural Polymorphism (Cullis and de Kruijff 1979)

Biological membranes are mainly composed of phospholipids and proteins in which molecular orientation in the hydrated membrane is determined by their

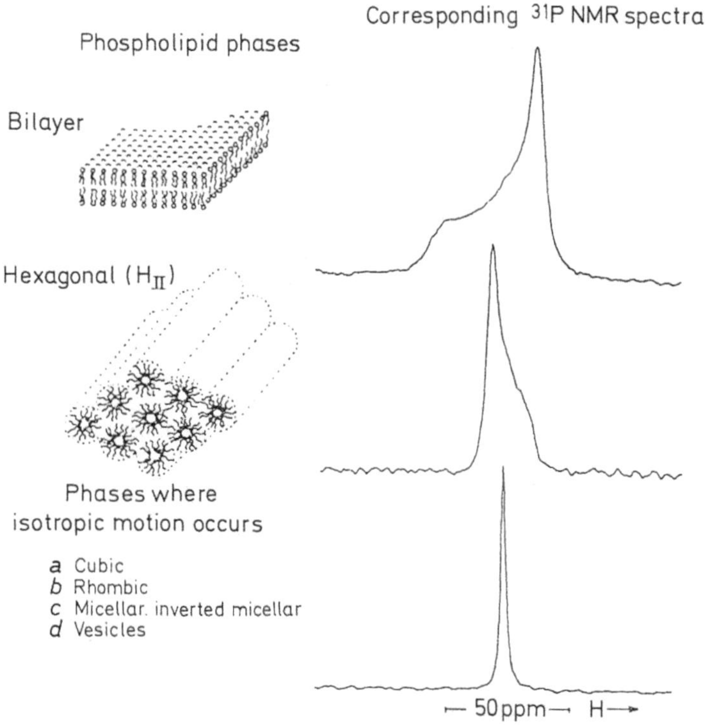

Fig. 12. Polymorphic phases available to hydrated liquid crystalline phospholipids and corresponding (36.4 MHz) ^{31}P-NMR spectra. (Cullis and de Kruijff 1979)

amphipathic nature: Phospholipids are oriented with polar heads outside the membrane towards the aqueous phases that lie on either side of the membrane; and apolar frame (hydrocarbon tails) inside the membrane. Hydrophobic parts of proteins are located inside the phospholipid bilayer, and hydrophilic parts emerge from the membrane. This lipid bilayer appears in a state of liquid crystal.

The introduction of ^{31}P-NMR methodology, which may be usefully applied to both models and biological membrane systems, has resulted in observations which imply that nonbilayer structures may occur in biological membrane (Cullis and de Kruijff 1979).

What are these structures? ^{31}P-NMR spectra of these membrane phases are realized in the presence of proton decoupling (Fig. 12).

Bilayer Phase: Rapid rotation of the phospholipids about their long axis results in a characteristic broad spectrum with a low field shoulder and high field peals. In biological membranes, the different phospholipids (PL) contribute to a composite bilayer lineshape.

Hexagonal Phase: H_{II}. In this structure, PL form long cylinders, with the polar head groups oriented into an aqueous core, additional motional averaging due to

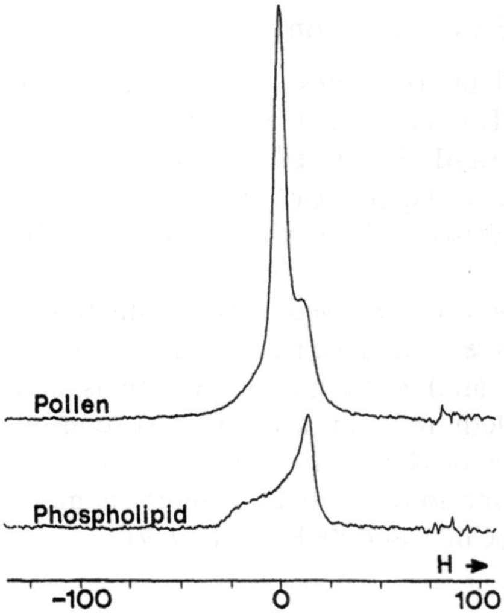

Fig. 13. ^{31}P-NMR spectra of fully hydrated. *T. latifolia* pollen and extracted phospholipid. The horizontal scale (chemical shift) is in ppm. The spurious signals in the 75 to 100 ppm range are instrumental artefacts. (Priestley and de Kruijff 1982)

lateral diffusion around aqueous channels, results in a ^{31}P-NMR lineshape asymmetry reversed compared to the bilayer spectra, and narrower by a factor of 2.

Micellar, Vesicule... Phase: In these configurations, PL undergo isotropic motions which result in a narrow, symmetrical ^{31}P-NMR spectrum.

These different configurations are easily pointed out with ^{31}P-NMR measurements in model systems, but not systems where ^{31}P-NMR saturation transfer techniques may be applied. This technique resolves the PL ^{31}P-NMR response from that of other phosphorus-containing compounds in intact tissue (de Kruijff et al. 1980b). Another technique for membrane configuration study results in the chemical extraction of PL (Priestley and de Kruijff 1982) (Fig. 13).

Nevertheless, this model system cannot really reflect living organization.

3.2 Functional Aspect of Membrane Polymorphism

3.2.1 Hexagonal Configuration: H_{II}

Most features favor bilayer to hexagonal transition: the nature of phospholipids (phosphatidylethanolamines prefer H_{II} configuration), unsaturated fatty acid, calcium, acidic pH, temperature, and dehydration (Cullis and de Kruijff 1979).

It is clear that this phase could not play a major role in biological membranes, as it is difficult to envisage such structures maintaining the permeability barrier vital to cellular integrity (see Simon 1978). The H_{II} configuration can only account for fusion between membranes when it appears in localized points of the membranes.

It is interesting to note that anhydrobiotic organisms inhibit this membrane transition in order to survive (Crowe et al. 1983). Probably, pollen membranes undergo H_{II} transition during dehydration. This results in an irreversible ametabolic state.

3.2.2 Vesicular, Micellar or Inverted Micellar Configuration

This configuration appears at physiological temperatures: in isolated microsomes, at 37 °C, while the majority of the PL undergo isotropic motion but at 4 °C, PL gives rise to the typical bilayer spectra (de Kruijff et al. 1980 a).

Again, there is no evidence for the existence of H_{II} phase of lipids in biological membrane, but ^{31}P-NMR signals indicating isotropic motion have been observed.

Thus, this nonbilayer configuration seems to be involved in many functional abilities of biological membranes as transbilayer transport (this accounts for localized nonbilayer configuration in the membrane). But this configuration is only an intermediate one: effectively, phosphatidylethanolamines adopt an H_{II} configuration. The addition of 15% mole of phosphatidylcholine (PC) induces an ^{31}P-NMR signal corresponding to an isotropic motion. When PC content is more than 30%, a bilayer arrangement is induced (Cullis and de Kruijff 1979).

3.2.3 Application to Water Stress

Crowe et al. (1983) have induced an isotropic signal on dry membranes by adding trehalose, and have so prevented a hexagonal transition of the membrane. Trehalose, a nonreducing disaccharide of glucose, is commonly found at high concentrations in anhydrobiotic organisms. Crowe et al. (1984) have studied trehalose (T) and PL interaction. They concluded that the addition of T to dry PL mimics addition of water. It is the "water replacement" hypothesis. During dehydration, the hydrogen bonds between T and PL may replace similar bonds between PL and water in bulk water.

Also, according to these authors, we can suggest that dry membranes, which present an isotropic signal in ^{31}P-NMR experiments, are water stress-resistant. In addition, we can conceive that this membrane is structured in vesicular configuration. In conclusion, modification of PL proportion during membrane turnover and the addition of a cryoprotectant such as trehalose could confer resistance to a dehydration-sensitive organism such as pumpkin pollen.

Evolution of membrane characteristics during such experiments could be examined with ^{31}P-NMR methodology or proton T_2 water. In fact, Clegg (1978) pointed out a progressive decrease of T_2, during *Artemia* cyst dehydration, followed at water content of 30% by an abrupt T_2 increasing. These data support the water replacement hypothesis.

3.3 ^{13}C-NMR and Membrane State

One of the fundamental properties of the lipid bilayer is its capacity to undergo a phase transition from a crystal gel to the liquid crystal state (Singer 1983). At low temperature, hydrated PL exists as a gel in which the hydrocarbon chains are closely packed and inclined relative to the membrane normal by an angle which is water content-dependent (Harrison and Lunt 1980). This T_2 transition is shown by DSC methodology (Becwar et al. 1983), but ^{13}C-NMR experiments could provide information on the motional state of the hydrocarbon chains of PL (Yeagle et al. 1978). This transition is associated with the alteration of the bilayer permeability and could account for freezing injury. It is thus well known that the membrane is the primary site of freezing injury (Levitt 1980; Pomeroy et al. 1983; Steponkus et al. 1983).

A modification of membrane fluidity by the introduction of foreign molecules, such as sterol, or by thermal acclimation (Hazel 1984) could overcome such injury.

4 Conclusions

The necessity to appreciate pollen quality arises in several contexts, including research on pollen-pistil recognition and interaction, in assessing the fertility of parent plants and hybrids in genetical experimentation, incompatibility phenomena and fertilization or during pollen storage and allergen identification. However, to be able to appreciate the quality of pollen, the need of complementary methods seems to be evident. As FCR is not a test of pollen quality, NMR methodology cannot determine precisely the right range of valuable measurements in this respect. Nevertheless, NMR provides an incredible amount of complementary information in terms of water content and membrane state. Because NMR results are statistic, we do not know if the loss of viability is a regular phenomenon during aging or if it corresponds to an instantaneous event for each grain. A specific signal from each pollen would be necessary to solve this question. In addition, the use of the so-called spin-imaging method could permit visualizing the hydric compartmentation evolution. This point is closely dependent on the NMR spectrometer sensibility. Theoretically 1 μg could be detectable, that is, a sphere of liquid water with a 62 μm diameter. In ^{31}P-NMR another limitation occurs in the case of partially hydrated pollen grain, because we cannot discriminate between membranes of different cellular compartments. Future developments may transform all empirical approaches to storing the pollen grains through a preliminary dehydration, and may avoid reaching the critical area where the vital water could be removed and the membrane denaturated.

Acknowledgments. We thank Mrs Miami for her technical assistance in NMR, Dr JC Duplan for helpful discussion, Mrs Frevre for reviewing the English and Ms AM Thierry for her laboratory assistance.

References

Barnabas B, Rajki E (1981) Fertility of deep frozen maize (*Zea mays* L.) pollen. Am J Bot 48:861–864

Becwar MR, Stanwood PC, Leonhardt KW (1983) Dehydration effects of freezing characteristics on survival in liquid nitrogen of desiccation-tolerant and desiccation-sensitive seeds. J Am Soc Hort Sci 108:613–617

Burke MJ, Gusta LV, Quamme HA, Weiser CJ, Li PH (1976) Freezing and injury plants. Annu Rev Plant Physiol 27:507–528
Clarke AE, Gleeson PA (1981) Molecular aspects of recognition and response in the pollen-stigma interaction. In: Loewus FA, Ryan CA (eds) The phytochemistry of cell recognition and cell surface interactions. Recent Adv Phytochem, vol. 15, Plenum, New York, pp 161–212
Clarke AE, Considine JA, Ward R, Knox RB (1977) Mechanism of pollination in *Gladiolus:* roles of the stigma and pollen tube guide. Nature 41:15–20
Clegg JS (1978) Hydration dependent metabolic transitions and the state of cellular water in Artemia. In: Crowe JH, Clegg JJ (eds) Dry biological systems. Academic, New York, pp 117–155
Clegg JJ (1979) Metabolism and the intracellular environment: the vicinal water network model. In: Drost Hansen W, Clegg JJ (eds) Cell associated water. Academic, New York, pp 363–413
Cohen JS (ed) (1983) Magnetic resonance in biology. Wiley Interscience, New York (vol. 2)
Crowe JH, Crowe LM, Mouradian R (1983) Stabilization of biological membranes at low water activities. Cryobiology 20:346–356
Crowe JH, Crowe LM, Chapman D (1984) Preservation of membranes in anhydrobiotic organism: the role of trehalose. Science 223:701–703
Cullis PR, Kruijff B de (1979) Lipid polymorphism and the functional roles of lipids in biological membranes. Biochim Biophys Acta 559:399–420
Drost Hansen W (1976) Structure and functional aspects of interfacial (vicinal) water. In: Alfsen A, Berteaud AJ (eds) L'eau et les systèmes biologiques. CNRS Paris, pp 177–186
Dumas C (1984) Water content and pollen physiology. In: Williams EG, Knox RB (eds) Pollination 84. University of Melbourne, pp 61–68
Dumas C, Gaude T (1983) Stigma-pollen recognition and pollen hydration. Phytomorphology 31:191–201
Dumas C, Duplan JC, Said C, Soulier JP (1983) ^{1}H nuclear magnetic resonance to correlate water content and pollen viability. In: Mulcahy PL, Ottaviano E (eds) Pollen: biology and implications for plant breeding. Elsevier Science, pp 15–20
Dumas C, Knox RB, Gaude T (1984a) Pollen recognition: new concepts from electron microscopy and cytochemistry. Int Rev Cytol 90:239–272
Dumas C, Duplan JC, Gaude T, Said C (1984b) Cytologie et physico-chimie, deux approches complémentaires pour tester la viabilité pollinique. In: INRA (ed) Ve Symp Int sur la pollinisation, pp 415–421
Dumas C, Knox RB, McConchie CA, Russell SD (1984d) Emerging physiological concepts in fertilization. What's news in plant physiol 15:17–20
Dumas C, Knox RB, Gaude T (1985) The spatial association of the sperm cells and vegetative nucleus in the pollen grain of *Brassica*. Protoplasma 124:168–174
Duplan JC, Dumas C (1984) Viabilité pollinique et conservation du pollen. In: Hervé Y, Dumas C (eds) Incompatibilité pollinique et amélioration des plantes. ENSA Rennes, pp 40–50
Etzler FM, Drost Hansen W (1979) A role for water in biological rate processes. In: Drost Hansen W, Clegg JJ (eds) Cell associated water. Academic, New York, pp 125–164
Gay G, Kerhoas C, Gaget M, Dumas C (1984) Basic aspects of pollen viability and storage. II Membrane state modifications. In: Ersam P (ed) VI Eur. Coll Cryobiology, 21–23 March 1984, Lyon, pp 156–160
Hamilton D (1976) Intersectional incompatibility in *Populus*. Ph D Thesis, Camberra
Harrison R, Lunt GG (eds) (1980) Biological membranes, their structure and function (2nd ed). Blackie Glasgow, London
Hazel JR (1984) Effects of temperature on the structure and metabolism of cell-membranes in fish. Am J Physiol 246:460–469
Hazlewood CF (1979) A view of the significance and understanding of the physical properties of cell-associated water. In: Drost Hansen W, Clegg JJ (eds) Cell associated water. Academic, New York, pp 165–260

Heslop-Harrison J (1975) Incompatibility and the pollen stigma interaction. Annu Rev Plant Physiol 26:403–425
Heslop-Harrison J (1979a) An interpretation of the hydrodynamics of pollen. Am J Bot 66:737–743
Heslop-Harrison J (1979b) Pollen walls as adaptative systems. Ann Mo Bot Gard 66:813–829
Heslop-Harrison J (1980) Compartmentation in anther development and pollen wall morphogenesis. In: Noverl, Eynene F, Nothe S (eds) Cell compartmentation and metabolic channeling
Heslop-Harrison J, Heslop-Harrison Y (1970) Evaluation of pollen viability by enzymatically induced fluorescence, intracellular hydrolysis of fluorescein diacetate. Stain Technol 45:115–120
Heslop-Harrison J, Knox RB, Heslop-Harrison Y, Mattson O (1975) Pollen-wall proteins: emission and role in incompatibility responses. In: Duckett JG, Racey PA (eds) The biology of the male gamete. Academic Press, London, pp 189–202
Heslop-Harrison J, Heslop-Harrison Y, Shivanna KR (1984) The evaluation of pollen quality, and a further appraisal of the fluorochromatic (FCR) test procedure. Theor Appl Genet 67:367–375
Heslop-Harrison Y, Shivanna KR (1977) The receptive surface of the angiosperm stigma. Ann Bot 41:1233–1258
Hesse M (1979) Development and ultrastructure of pollenkitt and exine in closely related entomophilous and anemophilous angiosperms: Salicaceae, Tiliaceae and Ericaceae. Flora 168:540–557
Hesse M (1980) Zur Stereostruktur der Viscinfäden bei Ericaceen und Onagraceen. Beitr Elektronenmikroskop Direktabb 13:211–216
Hoekstra FA (1984) Imbibitional chilling injury in pollen. Involvement of the respiratory chain. Plant Physiol 74:815–821
Howlett BJ, Knox RB (1984) Allergy interactions. In: Linskens HF, Heslop-Harrison J (eds) Cellular interactions. Springer, Berlin Heidelberg New York, pp 655–673 (Encyclopedia of plant physiology, New Series, vol 7)
Kerhoas C, Gay G, Duplan JC, Dumas C (in press) ^{1}H-NMR and pollen quality: correlation between T_2 relaxation time and fluorochromatic reaction (FCR) data. Proc Natl Acad Sci USA
Kerhoas C, Gay G, Gaget M, Vassoille R, Dumas C (1984) Basic aspects of pollen viability and storage. I Water content. In: Ersam P (ed) VI Eur Coll Cryobiology, 21–23 March, Lyon, pp 149–155
Knox RB (1984a) The pollen grain. In: Johri BM (ed) Embryology of angiosperms. Springer, Berlin Heidelberg New York, pp 1–98
Knox RB (1984b) Pollen-pistil interactions. In: Linskens HF, Heslop-Harrison (eds) Springer, Berlin Heidelberg New York, pp 508–592 (Encyclopedia of Plant Physiology, New Series, vol 17)
Knox RB, Heslop-Harrison J, Heslop-Harrison Y (1975) Pollen wall proteins: localization and characterization of gametophytic and sporophytic fractions. Biol J 7:177–187
Kruijff B de, Rietveld A, Cullis PR (1980a) P-NMR studies on membrane PL in microsomes, rat liver slices and intact perfused rat liver. Biochim Biophys Acta 600:343–357
Kruijff B de, Morris GA, Cullis PR (1980b) Application of ^{31}P-NMR saturation transfer techniques to investigate phospholipid motion and organization in model and biological membrane. Biochim Biophys Acta 598:206–211
Levitt J (1980) Responses of plants to environmental stresses. Levitt J (ed) Academic Press, New York
Mau SL, Raff J, Clarke AE (1982) Isolation and partial characterization of components of *Prunus avium* L. styles, including an antigenic glycoprotein associated with a self incompatibility genotype. Planta 156:505–516
Mazliak P (1983) Plant membrane lipids: changes and alterations during ageing and senescence. In: Lieberman M (ed) Post harvest physiology and crop preservation. Plenum, New York, pp 123–140

Mc Conchie Ca, Jobson S, Knox RB (1984) Analysis of the ultrastructure of sperm cells of *Brassica campestris* by computer assisted three dimensional reconstruction. In: Williams EG, Knox RB (eds) Pollination 84. University of Melbourne, pp 26–29

Peschel G, Belouschek P (1979) The problem of water structure in biological systems. In: Drost Hansen W, Clegg JJ (eds) Cell associated water. Academic, New York, pp 3–52

Pomeroy MK, Pihakaski SJ, Andrews CJ (1983) Membrane properties of isolated winter wheat cells in relation to icing stress. Plant Physiol 72:535–539

Porter EK (1981) Origins and gametic non variability of the proteins which diffuse from maize pollen. Environ Health Perspect 37:53–60

Priestley DA, Kruijff B de (1982) Phospholipid motional characteristics in a dry biological system. A ^{31}P-nuclear magnetic resonance study of hydrating *Typha latifolia* pollen. Plant Physiol 70:1075–1078

Raff J, Mc Kensie IFC, Clarke AE (1980) Antigenic determinants of *Prunus avium* are associated with the protoplast surface. Z Pflanzenphysiol 98:225–234

Ratkovic S, Bacic G, Radenovic E, Vucinic Z (1982) Water in plants. A review of some recent NMR studies concerning the state, and transport of water in leaf, root and seed. Stud Biophys 91:9–18

Roberts IN, Gaude T, Harrod G, Dickinson HG (1983) Pollen-stigma interactions in *Brassica oleracea:* a new pollen germination medium and its use in elucidating the mechanism of self-incompatibility. Theor Appl Genet 65:231–238

Rotman B, Papermaster BW (1966) Membrane porperties of living mammalian cells as studied by enzymatic hydrolysis of fluorogenic esters. Proc Natl Acad Sci USA 55:134–141

Russell SM (1983) Gametic fusion in *Plumbago zeylanica:* ultrastructural evidence for gamete recognition and the preferential transmission of sperm plastids into the zygote. Am J Bot 70:416–434

Simon EW (1978) Membranes in dry and imbibing seeds. In: Crowe JH, Clegg JJ (eds) Dry biological systems. Academic, New York, pp 205–224

Singer MA (1983) What have permeability experiments revealed about the structure of phospholipid bilayer membranes? In: Bangham AD (ed) Liposome letters. Academic, London, pp 105–112

Stanley RG, Linskens HF (1974) Pollen: biology, biochemistry, management. Springer, Berlin Heidelberg New York

Steponkus PL, Nowgert MF, Gormon-Kamm WJ (1983) Destabilization of the plasma membrane of isolated plant protoplasts during a freeze thaw cycle: the influence of cold acclimation. Cryobiology 20:448–465

Wiggins PM (1979) Metabolic control of the properties of intracellular water as a universal driving force for active transport. In: Drost Hansen W, Clegg JJ (eds) Cell associated water. Academic, New York, pp 69–114

Williams EG, Ramm Anderson S, Dumas C, Mau SL, Clarke AE (1982) The effect of isolated components of *Prunus avium* L. styles on in vitro growth of pollen tubes. Planta 156:517–519

Yeagle PL, Hutton WC, Martin RB (1978) Shingomyelin multiple phase behavior as revealed by multinuclear magnetic resonance spectroscopy. Biochemistry 17:5745–5750

Subject Index

MIX
Papier aus verantwortungsvollen Quellen
Paper from responsible sources
FSC® C105338

If you have any concerns about our products,
you can contact us on
ProductSafety@springernature.com

In case Publisher is established outside the EU,
the EU authorized representative is:
Springer Nature Customer Service Center GmbH
Europaplatz 3, 69115 Heidelberg, Germany

Printed by Libri Plureos GmbH
in Hamburg, Germany